ei **Addison-Wesley Series in Educational Innovation**

Newtonian Tasks Inspired by Physics Education Research

nTIPERs

Curtis J. Hieggelke
Joliet Junior College

David P. Maloney
Indiana University Purdue University Fort Wayne

Stephen E. Kanim
New Mexico State University

Addison-Wesley

Boston Columbus Indianapolis New York San Francisco Upper Saddle River
Amsterdam Cape Town Dubai London Madrid Milan Munich Paris Montréal Toronto
Delhi Mexico City São Paulo Sydney Hong Kong Seoul Singapore Taipei Tokyo

Publisher: Jim Smith

Senior Project Editor: Katie Conley

Editorial Assistant: Peter Alston

Managing Editor: Corinne Benson

Production Project Manager: Beth Collins

Cover Designer: Derek Bacchus

Manufacturing Buyer: Jeff Sargent

Senior Marketing Manager: Kerry Chapman

Cover Photo Credit: fotolia/Coco Brown

ISBN 10: 0-321-75375-5; ISBN 13: 978-0-321-75375-5

Addison-Wesley
is an imprint of

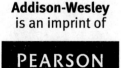

www.pearsonhighered.com

Addison-Wesley Series in Educational Innovation

Physlet® Quantum Mechanics: An Interactive Introduction
Mario Belloni
Wolfgang Christian
Anne J. Cox

Physlet® Physics: Interactive Illustrations, Explorations and Problems for Introductory Physics
Wolfgang Christian
Mario Belloni

Physlets®: Teaching Physics with Interactive Curricular Material
Wolfgang Christian
Mario Belloni

Peer Instruction: A User's Manual
Eric Mazur

Tutorials in Introductory Physics
Lillian C. McDermott
Peter S. Shaffer
The Physics Education Group, University of Washington

Just-In-Time Teaching: Blending Active Learning with Web Technology
Gregor M. Novak
Evelyn T. Patterson
Andrew D. Gavrin
Wolfgang Christian

Ranking Tasks in Physics: Student Edition
Curtis J. Hieggelke
David P. Maloney
Thomas L. O'Kuma

E&M TIPERs: Electricity and Magnetism Tasks
Curtis J. Hieggelke
David P. Maloney
Thomas L. O'Kuma
Stephen E. Kanim

About the Authors

Since 1991, Curtis Hieggelke, David Maloney, and Stephen Kanim have led over 30 workshops in which educators learned how to use and develop TIPERs. Many of these workshops were part of the Two Year College Physics Workshop Project (supported by seven grants from the National Science Foundation and co-directed by Curtis Hieggelke), which has offered a series of more than 60 professional development workshops for over 1200 Two-Year College and High School physics teacher participants. Working with Alan Van Heuvelen and Thomas O'Kuma, Maloney and Hieggelke also developed the Conceptual Survey of Electricity and Magnetism, which has become a standard instrument for measuring electricity and magnetism conceptual gains in introductory physics courses. The American Physical Society gave the 2009 Excellence in Physics Education Award to Hieggelke, Maloney, and O'Kuma in part for their work on TIPERs. This is the fourth book for Maloney and Hieggelke dealing with curriculum materials based on PER.

Curtis J. Hieggelke received a B.A. in physics and mathematics from Concordia College and a Ph.D. in theoretical particle physics from the University of Nebraska. He spent his professional career as a physics teacher at Joliet Junior College until he retired in 2003. Hieggelke has served as President and Section Representative of the Illinois Section of the American Association of Physics Teachers (AAPT), and received the Distinguished Service Citation in 1993. Joliet Junior College received the Illinois Community College Board Excellence in Teaching/Learning Award for his work in 1992. He was awarded the Distinguished Service Citation by AAPT in 1994 and he was elected to the Executive Board of AAPT for three years as the Two-Year College Representative. He has served as the Principal Investigator for 12 National Science Foundation (NSF) grants for workshops and curriculum materials development based on Physics Education Research (PER).

David P. Maloney received a B.S. in physics from the University of Louisville, and an M.S. in physics and Ph.D. in physics, geology, and education from Ohio University. A member of the Indiana University-Purdue University-Fort Wayne faculty since 1987, he has also taught at Wesleyan College and Creighton University. He was awarded the Distinguished Service Citation by AAPT in 2005. His main research interests concern the study of students' common sense ideas about physics, how those ideas interact with physics instruction, and the study of problem solving in physics. Maloney has authored or co-authored two-dozen articles and has been the Principal Investigator or Co-Principal Investigator for eight NSF grant projects.

Stephen E. Kanim received a B.S. in electrical engineering from UCLA and a Ph.D. in physics from the University of Washington, and has been a member of the physics faculty at New Mexico State University since 1998. He has varied research interests related to the teaching and learning of physics, including student conceptual understanding of specific physics topics, reasoning about proportions, and experimental tests of models of student thinking. He has worked on several research-based curriculum development projects including development of a set of introductory labs for mechanics. He previously taught high school physics in Las Cruces, New Mexico and in Palo Alto, California, and worked as an electrical engineer in Santa Clara, California.

CONTENTS

PREFACE

This workbook is intended to improve understanding of some of the ideas underlying Newtonian dynamics. As the subtitle of the workbook *Tasks Inspired by Physics Education Research* (TIPERs) suggests, the design of the individual tasks within the workbook is based on research that has been conducted into how students learn physics. This research has focused on students who are facing the challenges of an introductory physics course. Through interviews with these students and through analyses of their responses to examination and homework questions, physics education researchers are developing a better understanding of common difficulties with physics and of the types of exercises that foster improved understanding. In some cases, the tasks in this workbook are directly based on questions that researchers have used to probe student understanding. In other cases, we have written exercises that are not directly taken from research, but that focus on concepts that research suggests are challenging to students.

One general finding of physics education research is that many students in introductory physics courses who attend lectures, read the textbooks, and solve the homework problems still struggle with important physics concepts, principles, and relationships. This difficulty is sometimes a result of conflicts between common-sense ideas about how nature behaves (ideas based on everyday experiences) and the physics rules being learned (ideas based on rigorous investigation and controlled experimentation). Some of the tasks we have included in this workbook are intended to focus on these conflicts between everyday experience and how physicists think about concepts. By directly comparing these conflicting ways of thinking about specific concepts, and by discussing them with peers, we expect that students will clarify which of their initial ideas are useful and which need to be modified.

First and foremost one should understand that this is *not a book that is just read to get some needed information*. To benefit from this workbook, an individual must **actively** work with the tasks. The tasks in this book focus on conceptual understanding. Consequently, plugging numbers into a formula will seldom be the way to solve these tasks. Looking up an answer in a physics book is also *not* the way to deal with these tasks. Completing the tasks and then discussing answers with others is a useful way to benefit from this book. The more someone talks about these ideas and issues, the better the concepts will be learned and understood.

The tasks in this workbook will help deepen an individual's understanding of Newtonian dynamics. Beyond this, we would like these tasks to reinforce the sense that the *ideas* of science have coherence and power that extend beyond the facts and equations.

There are ten types of task formats in this book. These formats are likely to be new to most people so those people will require a little time to learn each format, that is, how the information is presented, and what someone needs to do to solve the task. The formats are reasonably straightforward, so getting familiar with them should not require significant time or effort. However, many of the tasks ask for explanations of the work or the reasoning associated with reaching the answer. This is one of the most important parts of these tasks. The explanation is especially important for cases where the correct answer is "none of them" or "it cannot be determined."

The different formats are designated by letters identifying the task type according to the following system: Bar Chart Tasks—BCT, Changing Representations Tasks—CRT, Comparison Tasks—CT, Conflicting Contentions Task—CCT, Linked Multiple Choice Tasks—LMCT, Qualitative Reasoning Tasks—QRT, Ranking Tasks—RT, Troubleshooting Tasks—TT, What, if anything, is Wrong Tasks— WWT, and Working Backwards Tasks—WBT. In addition to the format identifier, each task begins with an nT which indicates a Newtonian dynamics task. Task titles then describe the physical situation and finally identify the target quantity being asked about. It is unlikely that an instructor will go through this book in sequence; it is more likely that there will be some jumping around as an instructor chooses different formats to use at different times.

There are some unique aspects to consider for the Ranking Task format. If there are two, three, or four of the variations that have equivalent values for the target quantity, it is necessary to explicitly show that they are tied when writing the ranking sequence. For example, if choices A and C have the same ranking which is greater than B, then the answer would be AC in the first slot, followed by a blank slot, and then B would be in the last slot (or A, C, B can be put in sequence with a circle enclosing A and C). With ranking tasks it may not be possible to figure out specific numerical values for a quantity, but it may still be possible to compare the situations to decide which is largest and so on. Consequently, it is possible to rank the situations.

Several common conventions are employed in the tasks in this book. A circle with a dot in the center is used to represent a vector pointing out of the page, and a circle with an x in the center is used to represent a vector pointing into the page. Unless clearly stated, all grids have the same spacing and related drawings have the same scale. We ignore friction in these tasks unless explicitly identified.

ACKNOWLEDGMENTS

An endeavor like this book requires input from many people in addition to the authors. We sincerely thank the following individuals. At Joliet Junior College (Joliet, IL): Presidents J. D. Ross and Eugenia Proulx, Judy Bucciferro, Max Lee, Geoff White, and former student Leeanne Daoust. Glenn Westin and James Wolford kindly reviewed tasks during the initial development phase. These tasks were extensively reviewed and improved by Martha Lietz of Niles West High School (Skokie, IL), Dr. Robert Morse of St. Albans School (Washington, DC), and Dr. William P. Hogan (Joliet Junior College, IL). The contributions of our evaluator, Mel Sabella (Chicago State University), and formal field testers S. Bowen (Chicago State University), K. Coble (Chicago State University), D. Desbien (Estrella Mountain Community College, AZ), A. Escuardo (Harold Washington College, IL), E. Garcia (Chicago State University), T. Kuhn (Chicago State University), M. Lee (Joliet Junior College, IL), J. Milan (Harold Washington College, IL), T. O'Kuma (Lee College, TX), W. Waggoner (San Antonio College, TX), D. Wetli (Wake Technical Community College, NC) and D. Zoller (Olive Harvey College, IL) are deeply appreciated. We also valued the feedback from the many participants who have attended our workshops and the many other informal field testers from high schools, community colleges, and universities.

Since this book contains Tasks Inspired by Physics Education Research, we would like to acknowledge several of our most important inspirations. First, Alan Van Heuvelen deserves mention because his ALPS manual, which used bar charts and changing representation tasks, provided an early and strong guiding idea, and he was the co-developer of the Working Backwards (Jeopardy) format. Second, the University of Washington Physics Education Group frequently used the Conflicting Contentions format and provided ideas for issues in mechanics. Third, we thank the many PER investigators whose names we unfortunately cannot mention individually. Finally, we want to recognize and thank Arnold B. Arons and Paul G. Hewitt for their pioneering work in promoting conceptual understanding in physics courses.

We would like to acknowledge and thank the National Science Foundation (DUE #9952735 and #0125831) and Duncan McBride in particular, for supporting the development of these materials and the Physics Workshop Projects. Finally, we thank James Smith, Peter Alston, and the staff at Pearson for their assistance and their willingness to publish a unique type of book.

NTEX-RT1: STACKED BLOCKS—MASS OF STACK

Shown below are stacks of various blocks. All masses are given in the diagram in terms of M, the mass of the smallest block.

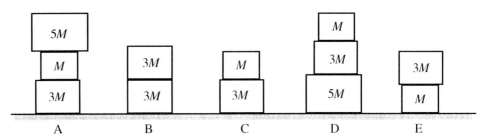

Rank the total mass of each stack.

　　　　　　Greatest　　1 _____ 2 _____ 3 _____ 4 _____ 5 _____ Least

OR, The magnitude of the total mass of each stack is the same but not zero. ____

OR, The magnitude of the total mass of each stack is zero. ____

OR, The ranking for the total mass of each stack cannot be determined. ____

Explain your reasoning.

Example answer formats
Stacks A and D have a total mass of 9M, C and E have a mass of 4M, and B has a mass of 6M or the ranking is A = D > B > C = E. Thus the ranking task answer should be expressed either as

　　　　　　Greatest　　1 ___AD__ 2 _____ 3 ___B___ 4 ___CE___ 5 _____ Least

or

　　　　　　Greatest　　1 ___A___ 2 ___D__ | 3 ___B___ | 4 ___C__ 5 ___E__ | Least

Note the order of equals is not important but it is easier if people are encouraged to use alphabetical order when possible.

An alternative but not preferred format is

　　　　　　Greatest　　1 ___AD__ 2 ___B___ 3 ___CE___ 4 _____ 5 _____ Least

NTPRACTICE-RT2: STACKED BLOCKS—NUMBER OF BLOCKS

Shown below are stacks of various blocks. All masses are given in the diagram in terms of M, the mass of the smallest block.

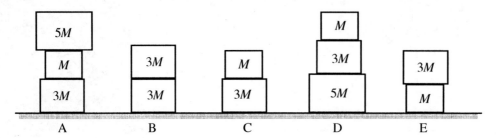

Rank the total number of blocks in each stack.

        Greatest   1 _____ 2 _____ 3 _____ 4 _____ 5 _____ Least

OR, The total number of is the same blocks for each stack.   ___

OR, The ranking for the total number of blocks in each stack cannot be determined.   ___

Explain your reasoning.

NTPRACTICE-RT3: STACKED BLOCKS—AVERAGE MASS

Shown below are stacks of various blocks. All masses are given in the diagram in terms of M, the mass of the smallest block.

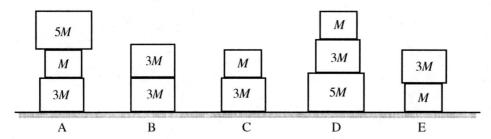

Rank the average mass in each stack.

        Greatest   1 _____ 2 _____ 3 _____ 4 _____ 5 _____ Least

OR, The average mass is the same for each stack.   ___

OR, The ranking for the average mass in these stacks cannot be determined.   ___

Explain your reasoning.

NT1A-RT1: CUTTING UP A BLOCK—DENSITY

A block of material (labeled A in the diagram) with a width w, height h, and thickness t, has a mass of M_o distributed uniformly throughout its volume. The block is then cut into three pieces, B, C, and D, as shown.

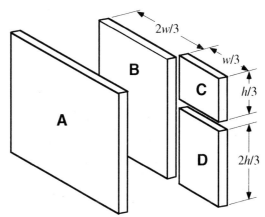

Rank the density of the original block A, piece B, piece C, and piece D.

Greatest   1 _____ 2 _____ 3 _____ 4 _____   Least

OR, The density is the same for all these pieces.   ____

OR, The ranking for the densities cannot be determined.   ____

Please explain your reasoning.

NT1A-WWT2: Cutting up a Block—Density

A block of material with a width w, height h, and thickness t, has a mass of M_o distributed uniformly throughout its volume. The block is then cut into two pieces, A and B, as shown. A student makes the following statement:

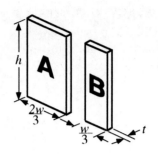

> *"The density is calculated by dividing the total mass by the volume. Since the volume is in the denominator, a large volume will give a small density. Therefore the block with the smallest volume, block B, will have the largest density."*

What, if anything, is wrong with the above statement? If something is wrong, explain the error and how to correct it. If the statement is correct, explain why.

NT1A-CCT3: Breaking up a Block—Density

A block of material with a width w, height h, and thickness t, has a mass of M_o distributed uniformly throughout its volume. The block is then broken into two pieces, A and B, as shown. Three students make the following statements:

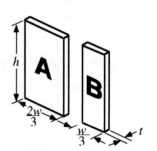

Andy: *"The density is the mass divided by the volume, and the volume of B is smaller. Since the mass is uniform and the volume is in the denominator, the density is larger for B."*

Badu: *"The density of piece A is larger than the density of piece B since A is larger, thus it has more mass."*

Coen: *"They both have the same density. It's still the same material."*

Which, if any, of these three students do you agree with?

Andy_____ Badu _____ Coen _____ None of them_____

Please explain your reasoning.

nT1A-QRT4: Slicing up a Block—Mass & Density

The block of material shown below has a length L_o and a volume V_o. An overall mass of M_o is spread uniformly throughout the volume of the block to give a density ρ_o and a linear density (in the direction of the measured length L_o) of λ_o.

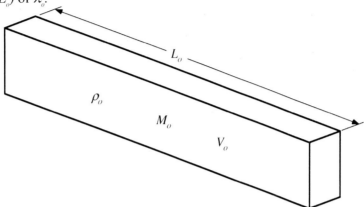

Three possible ways to slice the block into unequal pieces are shown below. In each case, the larger piece has a volume $2V_o/3$ and the smaller piece has a volume $V_o/3$.

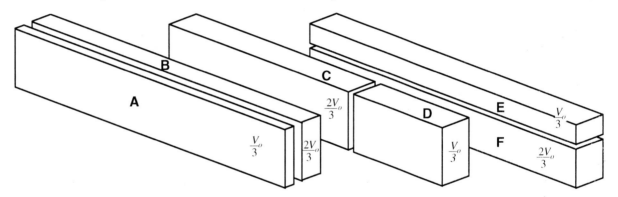

Fill in the table below with the quantities indicated for the pieces of the block labeled A – F in terms of the variables M_o, λ_o, and ρ_o.

	Mass	Mass per unit length	Mass per unit volume
Original block	M_o	λ_o	ρ_o
Piece A			
Piece B			
Piece C			
Piece D			
Piece E			
Piece F			

NT1A-QRT5: Cylindrical Rods with Same Mass—Volume, Area, and Density

Two cylindrical rods are shown. Rod A has a height H and a radius R and rod B has a height $2H$ and a radius $2R$. Both rods have the same total mass. Rod A has a density ρ_A and volume V_A.

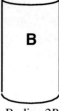

(a) What is the volume of rod B in terms of the volume of rod A?
(Your answer should look like $V_B = n\, V_A$, where n is some number.)

Please explain.

Radius R, Height H

Radius $2R$, Height $2H$

(b) What is the surface area of rod B in terms of the surface area of rod A? (Your answer should look like $SA_B = n\, SA_A$, where n is some number.)

Please explain.

(c) What is the density of Rod B in terms of the density of Rod A? (Your answer should look like $\rho_B = n\, \rho_A$, where n is some number.)

Please explain.

(d) What is the mass per unit length of Rod B to the mass per unit length of Rod A? (Your answer should look like $\lambda_B = n\, \lambda_A$, where λ is the mass per unit length and n is some number.)

Please explain.

nT1A-BCT6: FOUR BLOCKS—MASS AND DENSITY

The block of material shown to the right has a volume V_o. An overall mass M_o is spread evenly throughout the volume of the block so that the block has a uniform density ρ_o.

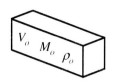

For each block shown below, the volume is given as well as *either* the mass or the density of the block.

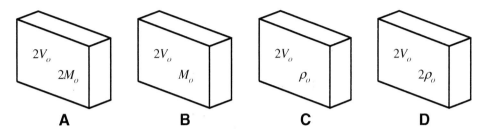

Construct two bar charts for the mass and density for the four blocks labeled **A – D** and for the pieces of the blocks if they were cut in half labeled **A/2 – D/2.** The mass and density for the original block is shown to set the scale of the chart.

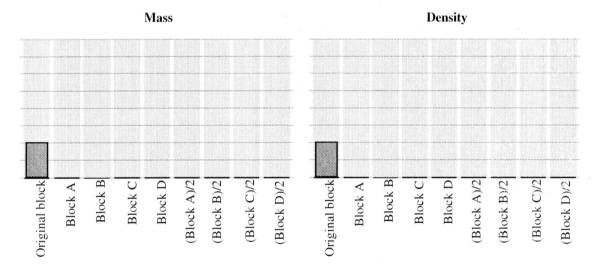

Please explain.

NT1B-CT7: SCALE MODEL PLANES—SURFACE AREA AND WEIGHT

A woodworker has made four small airplanes and one large airplane. All airplanes are exactly the same shape, and all are made from the same kind of wood. The larger plane is twice as large in every dimension as one of the smaller planes. The planes are to be painted and then shipped as gifts.

Case A   Case B

a) The amount of paint required to paint the planes is directly proportional to the surface area. **Will the amount of paint required for the single plane in Case A be *greater than, less than,* or *equal to* the total amount of paint required for all four planes in Case B?**

Please explain your reasoning.

b) The shipping cost for the planes is proportional to the weight. **Will the weight of the single plane in Case A be *greater than, less than,* or *equal to* the total weight of all four planes in Case B?**

Please explain your reasoning.

NT2A-QRT1: VECTORS ON A GRID I—MAGNITUDES

Eight vectors are shown below superimposed on a grid.

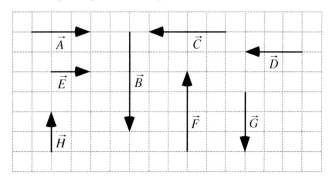

(a) List all of the vectors that have the same magnitude as vector $\vec{A}$.

(b) List all of the vectors that have the same magnitude as vector $\vec{B}$.

(c) List all of the vectors that have the same magnitude as vector $\vec{C}$.

(d) List all of the vectors that have the same magnitude as vector $\vec{D}$.

(e) List all of the vectors that have the same magnitude as vector $-\vec{A}$.

(f) List all of the vectors that have the same magnitude as vector $-\vec{B}$.

(g) List all of the vectors that have the same magnitude as vector $-\vec{C}$.

(h) List all of the vectors that have the same magnitude as vector $-\vec{D}$.

NT2A-RT2: VECTORS ON A GRID I—MAGNITUDES

Eight vectors are shown below superimposed on a grid.

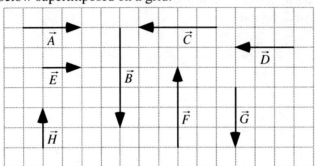

Rank the magnitudes of the vectors.

    Greatest 1_____ 2_____ 3_____ 4_____ 5_____ 6_____ 7_____ 8_____ Least

OR, All of these vectors have the same magnitude.   ____

OR, We cannot determine the ranking for the magnitudes of the vectors.   ____

Please explain your reasoning.

NT2A-QRT3: VECTORS ON A GRID II—DIRECTIONS

Nine vectors are shown below superimposed on a grid.

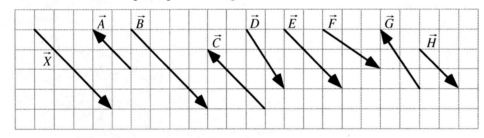

a) List all of the vectors that have the same direction as vector $\vec{X}$.

b) List all of the vectors that have the same direction as the vector $-\vec{X}$.

NT2B-CCT4: ADDING TWO VECTORS—MAGNITUDE OF THE RESULTANT

Three students are discussing the magnitude of the resultant of the addition of the vectors $\vec{A}$ and $\vec{B}$. Vector $\vec{A}$ has a magnitude of 5 centimeters, and vector $\vec{B}$ has a magnitude of 3 centimeters.

Alexis: *"We'd have to know the directions of the vectors to know how big the resultant is going to be."*

Bert: *"Since we are only asked about the magnitude, we don't have to worry about the directions. The magnitude is just the size, so to find the magnitude of the resultant we just have to add the sizes of the vectors. The magnitude of the resultant in this case is 8 centimeters."*

Cara: *"No, these are vectors, and to find the magnitude you have to use the Pythagorean theorem. In this case the magnitude is the square root of 34, a little less than 6 centimeters."*

Dacia: *"The resultant is the vector that you have to add to the first vector to get the second vector. In this case the resultant is 2."*

Which, if any, of these students do you agree with?

Alexis _____ Bert _____ Cara _____ Dacia_____ None of them_____

Please explain your reasoning.

NT2B-QRT5: VECTORS ON A GRID III—GRAPHICAL REPRESENTATION OF SUM

Shown below are four scaled vectors labeled $\vec{K}$, $\vec{L}$, $\vec{M}$, and $\vec{N}$ with lengths in arbitrary units.

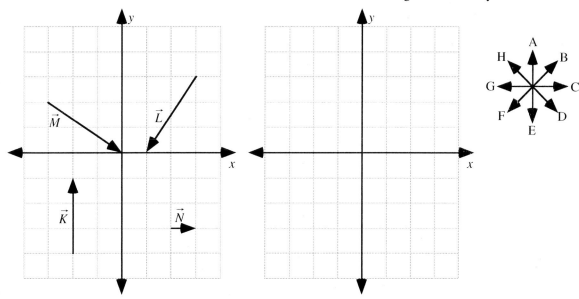

On the right, construct a graphical representation on the right of $\vec{J} = \vec{K} + \vec{L} + \vec{M} + \vec{N}$ with labels for each vector, and indicate the direction of $\vec{J}$ _____ (closest to one of the directions listed in the direction rosette above).

NT2B-RT6: Vectors I—Resultant Magnitudes of Adding Two Vectors

Eight vectors are shown superimposed on a grid.

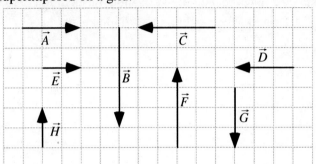

Rank the magnitude of the vector resulting from adding vector $\vec{A}$ to each vector ($\vec{A}+\vec{A}$, $\vec{B}+\vec{A}$, $\vec{C}+\vec{A}$, etc).

    Greatest 1_____ 2_____ 3_____ 4_____ 5_____ 6_____ 7_____ 8_____ Least

OR, All of these resulting vectors have the same magnitude.   ____

OR, We cannot determine the ranking for the magnitudes of the resulting vectors.   ____

Please explain your reasoning.

NT2B-CCT7: COMBINING TWO VECTORS—RESULTANT

Two vectors each have a magnitude 6 units, and each makes a small angle α with the horizontal as shown. Four students are arguing about the resultant vector obtained by adding these two vectors.

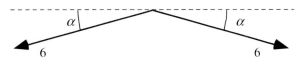

Amanda: *"Since these are vectors, we need to use the Pythagorean theorem to find the magnitude. In this case, the magnitude will be the square root of 72. The direction will be downward."*

Belle: *"Since these are vectors we have to find a direction and a magnitude. We use the vectors to determine the direction, which is down. But to get the magnitude, we just add the individual magnitudes. The magnitude of the resultant is twelve."*

Conrad: *"I think the direction is actually up. The resultant should add to these vectors to get zero, and since these ones point down, we need another vector pointing up."*

Donald: *"The magnitude will be less than six. Each of these point down just a little so the resultant will be pretty small."*

Which, if any, of these students do you agree with?

Amanda _____ Belle _____ Conrad _____ Donald _____ None of them _____

Please explain your reasoning.

NT2B-CT8: COMBINING VECTORS—MAGNITUDE OF RESULTANT

In Case A, two vectors of magnitude 6 units are at right angles to one another. In Case B, four vectors, each of magnitude 3 units, are arranged as shown. The outer vectors in Case B are also at right angles to one another, and the difference in direction between any pair of adjacent vectors is 30°.

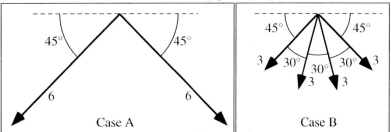

If all vectors in each case are added together, is the magnitude of the resultant in Case A *greater than, less than,* or *equal to* the magnitude of the resultant in Case B?

Please explain your reasoning.

NT2B-QRT9: VECTOR COMBINATION II—DIRECTION OF RESULTANT

For each situation below, combine the vectors as indicated and determine the direction of the resultant vector. Then select the closest direction to the resultant from the direction rosette at the right.

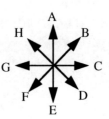

$\vec{J} = \vec{N} + \vec{P} + \vec{Q} + \vec{R}$

Direction of $\vec{J}$: _____

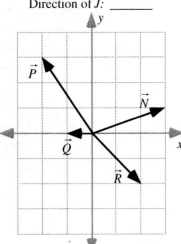

$\vec{K} = \vec{P} + \vec{R} + \vec{N} + \vec{Q}$

Direction of $\vec{K}$: _____

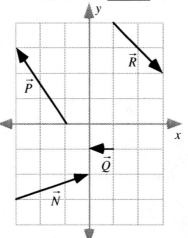

$\vec{L} = -\vec{S} - \vec{T} + \vec{U} - \vec{V}$

Direction of $\vec{L}$: _____

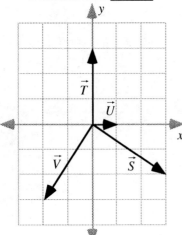

$\vec{M} = \vec{U} - \vec{T} - \vec{V} - \vec{S}$

Direction of $\vec{M}$: _____

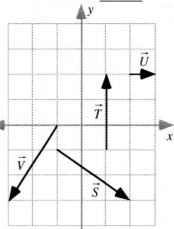

nT2C-CCT10: Two Vectors—Vector Difference

Two vectors labeled $\vec{A}$ and $\vec{B}$, each having a length of 6 meters, make a small angle α with the horizontal as shown. Four students are arguing about the vector difference $\vec{C} = \vec{A} - \vec{B}$.

Arlo: *"Since we're subtracting vector B, we flip it around so it points in the same direction as vector A. The difference will be 12 meters long and will point in the same direction as vector A."*

Bob: *"We're subtracting, so the resultant will be smaller than six. Both vectors point down, so the difference will point down as well."*

Celine: *"When you flip vector B around to get negative B, it points up and to the left. Then we add it to vector A, we get a long vector pointing horizontally to the right."*

Delbert: *"Both vectors are 6 meters long, so the difference is zero. It doesn't point in any direction."*

Which, if any, of these students do you agree with?

Arlo _____ Bob _____ Celine _____ Delbert _____ None of them _____

Please explain your reasoning.

nT2C-CT11: Two Vectors—Vector Sum and Difference

Two vectors labeled $\vec{A}$ and $\vec{B}$ each have a magnitude of 6 meters, and each makes a small angle α with the horizontal as shown. Let $\vec{C} = \vec{A} + \vec{B}$ and $\vec{D} = \vec{A} - \vec{B}$.

Is the magnitude of $\vec{C}$ *greater than*, *less than*, or *equal to* the magnitude of $\vec{D}$?

Please explain your reasoning.

NT2C-RT12: Addition and Subtraction of Three Vectors I—Magnitude

Shown are three vectors labeled $\vec{R}$, $\vec{S}$, and $\vec{T}$, with lengths given in arbitrary units.

Rank the magnitudes of the vectors $\vec{A}$ through $\vec{F}$ formed by adding and subtracting vectors $\vec{R}$, $\vec{S}$, and $\vec{T}$ as described below.

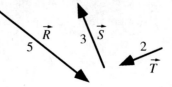

$$\vec{A} = \vec{R} + \vec{S} + \vec{T} \qquad\qquad \vec{B} = \vec{R} - \vec{S} + \vec{T} \qquad\qquad \vec{C} = \vec{T} + \vec{R} - \vec{S}$$

$$\vec{D} = \vec{R} - \vec{S} - \vec{T} \qquad\qquad \vec{E} = \vec{S} + \vec{T} - \vec{R} \qquad\qquad \vec{F} = \vec{R} + \vec{S} - \vec{T}$$

Greatest   1 _____   2 _____   3 _____   4 _____   5 _____   6 _____   Least

OR, The magnitude of the vectors $\vec{A} - \vec{F}$ is the same.   ___

OR, The ranking for the magnitude of the vectors $\vec{A} - \vec{F}$ cannot be determined.   ___

Please explain your reasoning.

NT2C-RT13: ADDITION AND SUBTRACTION OF THREE VECTORS II—DIRECTION OF RESULTANT

Three vectors, labeled $\vec{P}$, $\vec{Q}$, and $\vec{R}$, are shown below. The magnitude of each vector is given in arbitrary units.

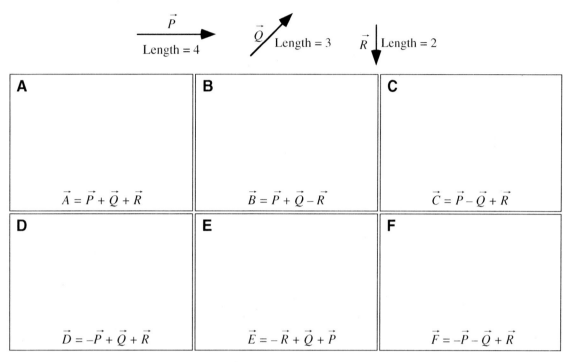

In each space provided above, construct a drawing of the resultant of the combination of the vectors $\vec{P}$, $\vec{Q}$, and $\vec{R}$, as indicated and then rank the magnitude of the angle that the resultant vector makes with the vector $\vec{P}$.

Greatest 1 _____ 2 _____ 3 _____ 4 _____ 5 _____ 6 _____ Least

Please explain your reasoning.

NT2C-RT14: ADDITION AND SUBTRACTION OF THREE VECTORS II—MAGNITUDE OF RESULTANT

Three vectors, labeled $\vec{P}$, $\vec{Q}$, and $\vec{R}$, are shown below. The magnitude of each vector is given in arbitrary units.

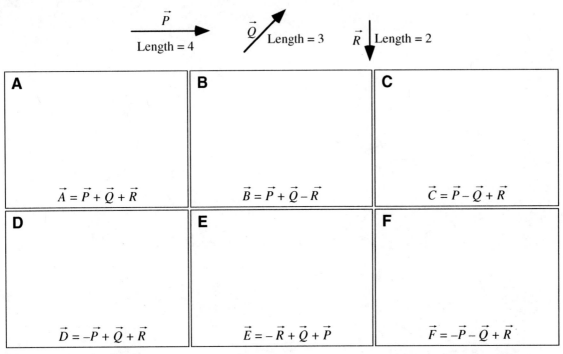

A	B	C
$\vec{A} = \vec{P} + \vec{Q} + \vec{R}$	$\vec{B} = \vec{P} + \vec{Q} - \vec{R}$	$\vec{C} = \vec{P} - \vec{Q} + \vec{R}$

D	E	F
$\vec{D} = -\vec{P} + \vec{Q} + \vec{R}$	$\vec{E} = -\vec{R} + \vec{Q} + \vec{P}$	$\vec{F} = -\vec{P} - \vec{Q} + \vec{R}$

In each space provided above, construct a drawing of the resultant of the combination of the vectors $\vec{P}$, $\vec{Q}$, and $\vec{R}$ as indicated and then rank the magnitude of these resultant vectors.

Greatest   1 _____ 2 _____ 3 _____ 4 _____ 5 _____ 6 _____ Least

OR, We cannot determine the ranking for the magnitude of the resultant vectors.   ___

Please explain your reasoning.

nT2D-QRT15: VECTOR COMBINATIONS III—COMPONENTS OF THE RESULTANT VECTOR

For each situation below, determine the components of the resultant vectors.

$\vec{K} = \vec{A} + \vec{B} + \vec{C} + \vec{D}$

$\vec{K}$: x-component _____

$\vec{K}$: y-component _____

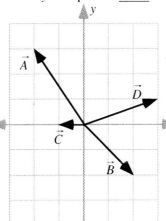

$\vec{L} = -\vec{E} - \vec{F} + \vec{G} + \vec{H}$

$\vec{L}$: x-component _____

$\vec{L}$: y-component _____

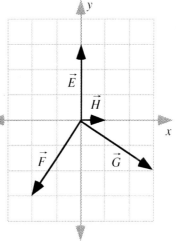

$\vec{M} = \vec{A} + \vec{B} + \vec{C} + \vec{D}$

$\vec{M}$: x-component _____

$\vec{M}$: y-component _____

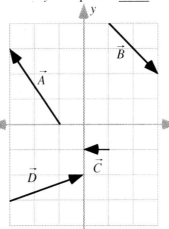

$\vec{N} = \vec{E} - \vec{F} - \vec{G} + \vec{H}$

$\vec{N}$: x-component _____

$\vec{N}$: y-component _____

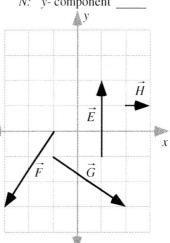

NT2D-QRT16: FORCE VECTORS—PROPERTIES OF COMPONENTS

Shown below are vector diagrams representing two sets of forces.

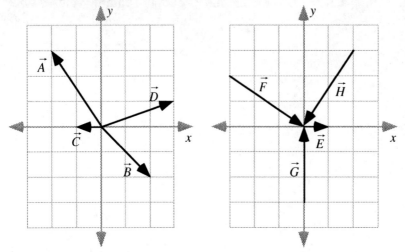

1. List all the forces that have a zero x-component:

2. List all the forces that have a zero y-component:

3. List all the forces that have an x-component pointing in the positive x-direction:

4. List all the forces that have a y-component pointing in the negative y-direction:

NT2D-QRT17: VELOCITY VECTORS—PROPERTIES OF COMPONENTS

Shown below are vector diagrams representing velocities.

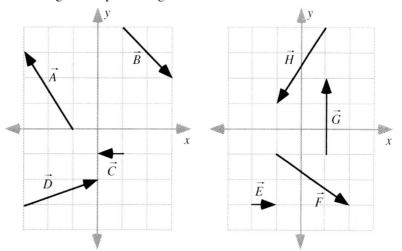

1. List all the velocities that have a zero *x*-component:

2. List all the velocities that have a zero *y*-component:

3. List all the velocities that have an *x*-component pointing in the positive *x*-direction:

4. List all the velocities that have a *y*-component pointing in the negative *y*-direction:

NT2D-CCT18: VECTOR—RESOLUTION INTO COMPONENTS

Three students are looking at three different solutions to a problem that includes the resolution of vector $\vec{A}$ into components as shown.

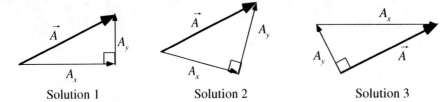

| Solution 1 | Solution 2 | Solution 3 |

Ayesha: *"Only Solution 1 is right. When we resolve a vector, we have to break it up into its x- and y-components so that the components add up to the original vector."*

Bashir: *"Solution 3 is right as well. The components still add to vector A, and the vectors still form a right triangle."*

Claudio: *"All three are right. The only difference is the choice of coordinate axes, and that will depend on what is most convenient for the problem solution."*

Which, if any, of these students do you agree with?

Ayesha _____ Bashir _____ Claudio _____ None of them_____

Please explain your reasoning.

NT2D-CT19: VECTOR ON ROTATED AXES—COMPONENTS

Shown below is a vector $\vec{F}$ and two sets of axes that are at some angle relative to each other.

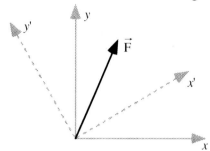

(a) Is the magnitude of $\vec{F}_x$ (the x-component of $\vec{F}$) *greater than, less than,* or *equal to* $\vec{F}_{x'}$ (the x'-component of $\vec{F}$)?
Explain.

(b) Is the magnitude of $\vec{F}_y$ (the y-component of $\vec{F}$) *greater than, less than,* or *equal to* $\vec{F}_{y'}$ (the y'-component of $\vec{F}$)?
Explain.

NT2D-CCT20: VECTOR COMPONENTS—RESULTANT VECTORS

Two vectors have components (A_x, A_y) and (B_x, B_y) where A_x is equal to B_y and A_y is equal to B_x. Three students make the following contentions:

Antonio: *"I think since these two vectors have components that are equal to each other the two vectors have to be equal also."*

Benito: *"No, I don't think we can say anything about how these two vectors compare because the same components are not equal for the two of them."*

Carlito: *"Well I disagree with both of you. The two vectors do have the same magnitude, but they do not have the same direction."*

Which, if any, of these three students do you agree with and think is correct?

Antonio _____ Benito _____ Carlito _____ None of them _____
Please explain your reasoning.

NT2E-WBT21: CALCULATIONS WITH FOUR VECTORS—VECTOR OPERATION

Students are given the following vectors with arbitrary units:

$\vec{A} = 6\hat{i} + 2\hat{j} = 2\sqrt{10}$ @ 18.435°

$\vec{B} = -4\hat{i} - 7\hat{j} = \sqrt{65}$ @ 240.3°

$\vec{C} = 19.28\hat{i} + 22.98\hat{j} = 30$ @ 50°

$\vec{D} = -38.64\hat{i} - 10.35\hat{j} = 40$ at 195°

a. Using the vectors above, what vector operation results in the following calculation?

$$(-24 - 14) \text{ units}^2 = -38 \text{ units}^2$$

b. Using the vectors above, what vector operation results in the following calculation?

$$= |30| \cdot |40| \cdot \cos(195° - 50°) = 1200 \cdot \cos(145°) = 1200 \cdot (-0.81915) = -983.0 \text{ units}^2$$

c. Using the vectors above, what vector operation results in the following calculation?

$$= |30| \cdot |40| \cdot \sin(195° - 50°) = 1200 \cdot \sin(145°) = 688.3 \text{ units}^2$$

d. Using the vectors above, what vector operation results in the following calculation?

$$= \begin{vmatrix} \hat{i} & \hat{j} & \hat{k} \\ +6 & +2 & 0 \\ -4 & -7 & 0 \end{vmatrix} = \hat{k} \begin{vmatrix} 6 & 2 \\ -4 & -7 \end{vmatrix} = (-42 + 8)\hat{k} = -34\hat{k} \text{ units}^2$$

NT2E-WBT22: Vectors on a Grid—Scalar (Dot) Product Expression

Four vectors are shown below superimposed on a grid that has arbitrary units.

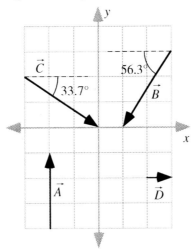

a. Using the vectors on the grid above, what scalar (dot) product results in the following calculation?

$$= \left|3 \text{ units}\right| \cdot \left|\sqrt{13} \text{ units}\right| \cdot \cos(146.3°) \text{ or } (3\hat{j}) \bullet (-2\hat{i} - 3\hat{j}) = (0-9) \text{ units}^2 = -9 \text{ units}^2$$

b. Using the vectors on the grid above, what scalar (dot) product results in the following calculation?

$$= \left|3 \text{ units}\right| \cdot \left|\sqrt{13} \text{ units}\right| \cdot \cos(123.7°) \text{ or } (3 \text{ units } \hat{j}) \bullet (3 \text{ units } \hat{i} - 2 \text{ units } \hat{j}) = (0-6) \text{ units}^2 = -6 \text{ units}^2$$

c. Using the vectors on the grid above, what scalar (dot) product results in the following calculation?

$$= \left|3 \text{ units}\right| \cdot \left|1 \text{ unit}\right| \cdot \cos(90°)$$

d. Using the vectors on the grid above, what scalar (dot) product results in the following calculation?

$$= \left|\sqrt{13} \text{ units}\right| \cdot \left|\sqrt{13} \text{ units}\right| \cdot \cos(90°) \text{ or } (3 \text{ units } \hat{i} - 2 \text{ units } \hat{j}) \bullet (-2 \text{ units } \hat{i} - 3 \text{ units } \hat{j}) = (-6+6) \text{ units}^2 = 0$$

e. Using the vectors on the grid above, what scalar (dot) product results in the following calculation?

$$= (3 \text{ units } \hat{i} - 2 \text{ units } \hat{j}) \bullet (1 \text{ unit } \hat{i}) = (3+0) \text{ units}^2 = 3 \text{ units}^2$$

f. Using the vectors on the grid above, what scalar (dot) product results in the following calculation?

$$= \left|\sqrt{13} \text{ units}\right| \cdot \left|1 \text{ unit}\right| \cdot \cos(33.7°)$$

NT2E-WBT23: Vectors on a Grid—Vector (Cross) Product

Four vectors are shown below superimposed on a grid that has arbitrary units.

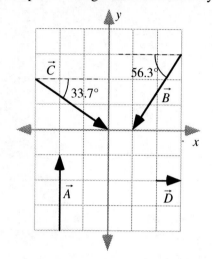

a) **Using the vectors on the grid above, what vector (cross) product results in the following calculation?**

$$= \begin{vmatrix} \hat{i} & \hat{j} & \hat{k} \\ 3 & -2 & 0 \\ -2 & -3 & 0 \end{vmatrix} = \hat{k} \begin{vmatrix} 3 & -2 \\ -2 & -3 \end{vmatrix} = (-9-4)\hat{k} = -13\hat{k}$$

b) **Using the vectors on the grid above, what vector (cross) product results in the following calculation?**

$$= \begin{vmatrix} \hat{i} & \hat{j} & \hat{k} \\ 0 & +3 & 0 \\ +1 & 0 & 0 \end{vmatrix} = \hat{k} \begin{vmatrix} 0 & +3 \\ +1 & 0 \end{vmatrix} = -3\hat{k}$$

NT2E-QRT24: VECTORS ON A GRID—PRODUCT EXPRESSIONS THAT ARE ZERO

Four vectors are shown below superimposed on a grid that has arbitrary units.

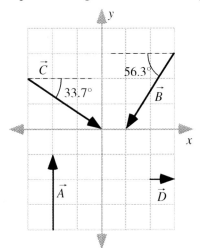

a) **Using the vectors on the grid above, list any scalar (dot) products of two of these vectors that are equal to zero.**

   Explain.

b) **Using the vectors on the grid above, list any vector (cross) products of two of these vectors that are equal to zero.**

   Explain.

NT3A-WWT1: VELOCITY VS. TIME GRAPH II—ACCELERATION VS. TIME GRAPH

A student obtains a graph of an object's velocity versus time and then draws the graph of the acceleration versus time for the same time interval.

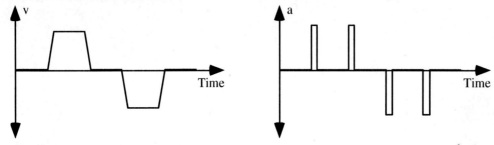

What, if anything, is wrong with the graph of the acceleration versus time? If something is wrong, identify it and explain how to correct it. If the graph is correct, explain why.

NT3A-CT2: VELOCITY VS. TIME GRAPHS—DISPLACEMENT

The graphs represent the velocity of two toy robots moving in one dimension for a particular time interval. Both graphs have the same time and velocity scales.

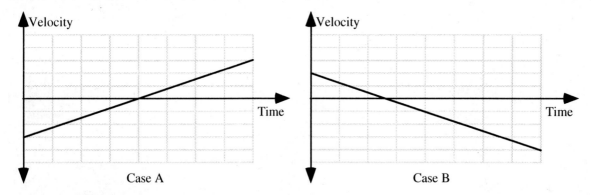

Is the magnitude of the displacement of the robot in Case A *greater than, less than,* or *equal to* the magnitude of the displacement of the robot in Case B?

Please explain your reasoning.

nT3A-WWT3: Acceleration vs. Time Graph—Velocity vs. Time Graph

A student obtains a graph of an object's acceleration versus time and then draws the graph of the velocity versus time for the same time interval. The object starts from rest.

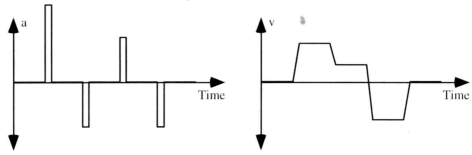

What, if anything, is wrong with the graph of velocity versus time? If something is wrong, identify it and explain how to correct it. If the graph is correct, explain why.

nT3A-WWT4: Velocity vs. Time Graph—Acceleration vs. Time Graph

A student obtains a graph of an object's velocity versus time and then draws the graph of the acceleration versus time for the same time interval.

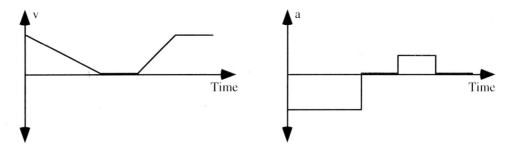

What, if anything, is wrong with the graph of the acceleration versus time? If something is wrong, identify it and explain how to correct it. If the graph is correct, explain why.

NT3A-CRT5: ACCELERATION VS. TIME GRAPH—VELOCITY VS. TIME GRAPH

Sketch a possible velocity versus time graph given the acceleration graph for the same time interval.

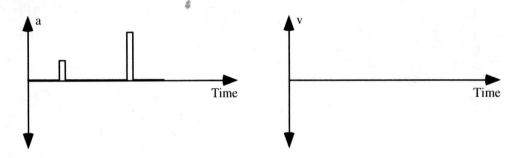

Explain.

NT3A-CRT6: VELOCITY VS. TIME GRAPH—ACCELERATION VS. TIME GRAPH

Sketch the acceleration versus time graph given the velocity versus time graph for the same time interval.

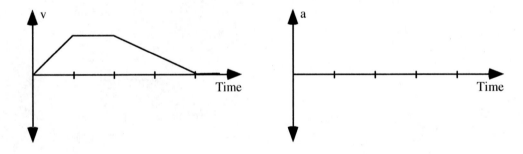

Explain.

NT3A-CT7: VELOCITY VS. TIME GRAPHS OF TWO OBJECTS I—DISPLACEMENT

The graphs below show the velocity of two objects during the same time interval.

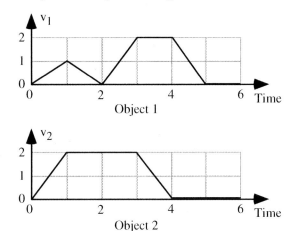

After 5 seconds, is the displacement of Object 1 in the top graph *greater than, equal to,* or *less than* the displacement of Object 2 in the bottom graph?

Please explain.

NT3A-CT8: VELOCITY VS. TIME GRAPHS OF TWO OBJECTS II—DISPLACEMENT

The graphs below show the velocity of two objects during the same time interval.

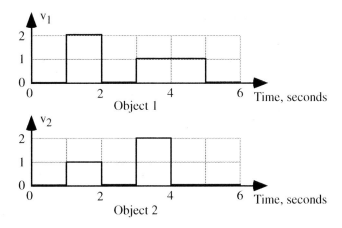

After 5 seconds, is the displacement of Object 1 in the upper graph *greater than, equal to,* or *less than* the displacement of Object 2 in the lower graph?

Please explain.

NT3A-RT9: VELOCITY VS. TIME GRAPHS—DISPLACEMENT

Shown below are six velocity-time graphs for toy robots that are traveling along a straight course. All the robots are initially facing the same way. All graphs have the same time and velocity scales.

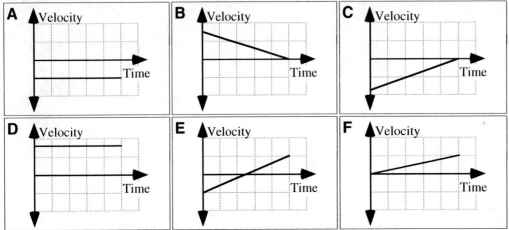

Rank these situations on the basis of the displacement during these intervals.

Greatest   1 _____ 2 _____ 3 _____ 4 _____ 5 _____ 6 _____ Least

OR, The displacement during these intervals is the same but not zero for all these robots.   ___

OR, The displacement during these intervals is zero for all these robots.   ___

OR, We cannot determine the ranking for the displacements of these robots.   ___

Please explain your reasoning.

NT3A-WWT10: BALL THROWN UPWARD AND COMES BACK DOWN—VELOCITY VS. TIME GRAPH

A ball is thrown straight upward and falls back to the same height. A student makes this graph of the velocity of the ball as a function of time.

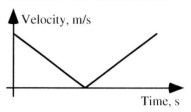

What, if anything, is wrong with the student's graph? If something is wrong, explain the error and how to correct it. If the graph is correct, explain why.

NT3A-WWT11: VELOCITY VS. TIME GRAPH OF TWO OBJECTS—FASTEST OBJECT

A student is shown the velocity-time graphs for two objects and is asked to decide which object is moving faster. The student responds:

"B is faster because it has the steeper slope."

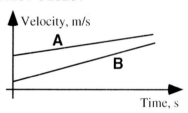

What, if anything, is wrong with the student's statement? If something is wrong, explain the error and how to correct it. If the statement is correct, explain why.

NT3A-CRT12: TRAVELING STUDENTS—VELOCITY VS. TIME GRAPH

Carmela and Desi leave their physics classroom separately and travel west. They both start from rest. Desi left first, traveling with an acceleration of 4 m/s^2 west for the first 6 seconds, and then he traveled at a constant velocity. Two seconds after Desi started, Carmela began with an acceleration of 3 m/s^2 west for 10 seconds, and after that she traveled at a constant velocity.

Graph the velocity of both travelers as a function of time up to $t = 16$ seconds starting at time $t = 0$ when Desi leaves the classroom. Use a solid line for Desi's velocity and a dashed line for Carmela's velocity.

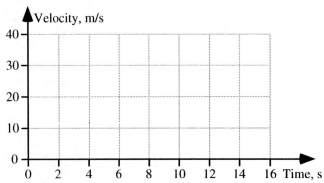

NT3A-CCT13: VELOCITY VS. TIME GRAPH—DISPLACEMENT

The graph shown represents the velocity of a toy robot moving in one dimension for a particular time interval. Three students studying this graph make the following statements:

Arnold: *"The robot's displacement is positive because the slope of the graph is positive."*

Betty: *"No, the robot's displacement will be zero since it moves in both the positive and negative directions during this time."*

Cindy: *"I think the displacement is negative since the robot has a negative velocity for a longer time."*

Which, if any, of these three students do you agree with and think is correct?

Arnold _____ Betty _____ Cindy _____ None of them_____

Please explain your reasoning.

NT3A-RT14: Velocity vs. Time Graphs—Distance Traveled

Velocity-time graphs for six toy robots that are traveling along a straight path are shown. All graphs have the same time and velocity scales.

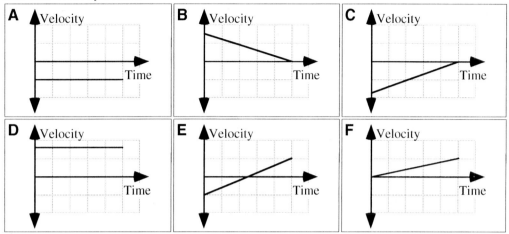

Rank these situations on the basis of the distance traveled during these intervals.

Greatest   1 _____ 2 _____ 3 _____ 4 _____ 5 _____ 6 _____ Least

OR, The distance during the intervals indicated is the same but not zero for all these robots.   ____

OR, The distance during the intervals indicated is zero for all these robots.   ____

OR, We cannot determine the ranking for the distances traveled of these robots.   ____

Please explain your reasoning.

NT3A-BCT15: Velocity vs. Time Graph—Distance Traveled

The graph represents the motion of a toy robot moving in one dimension during a 14-second interval.

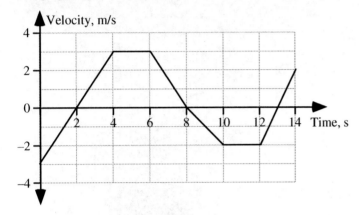

In the histogram below, the bar represents the distance the robot travels during the 2-second interval from 4 s to 6 s. **Draw additional bars to represent the distance traveled during the other 2-second intervals.**

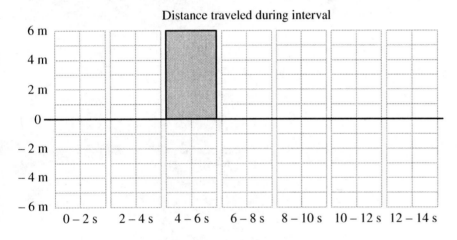

Distance traveled during interval

NT3A-CCT16: VELOCITY VS. TIME GRAPHS OF TWO OBJECTS I—DISPLACEMENT

The graphs below show the velocity of two objects during the same time interval.

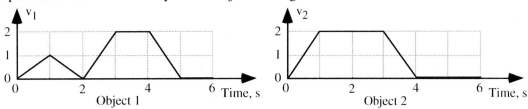

Three students are discussing the displacements of these objects for this interval.

Andy: *"I think Object 2 will have the greater displacement because it gets to a higher speed faster than Object 1."*

Badu: *"No, Object 1 will have the greater displacement because it travels for longer than Object 2."*

Connor: *"I think Andy has the right answer, but for the wrong reason. We can see that Object 2 has the larger displacement because the area under the graph is greater."*

Which, if any, of these three students do you agree with?

Andy_____ Badu _____ Connor _____ None of them_____

Please explain your reasoning.

NT3A-CCT17: VELOCITY VS. TIME GRAPHS OF TWO OBJECTS II—DISPLACEMENT

The graphs below show the velocity of two objects during the same time interval.

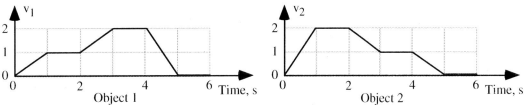

Three students are discussing the displacements of these objects for this interval.

Apriel: *"I think Object 2 will have the greater displacement because it gets to a higher speed faster than Object 1."*

Brody: *"Object 1 spends most of its time speeding up, but object 2 spends most of its time slowing down. Object 1 will go farther."*

Cyril: *"The displacement is found from the integral or area of the velocity graphs. But in this case we don't know what the integration constant or the initial position is that we need to add to the integral or area. We don't have enough information to find the displacement."*

Which, if any, of these three students do you agree with?

Apriel_____ Brody _____ Cyril _____ None of them_____

Please explain your reasoning.

NT3A-CRT18: Velocity vs. Time Graphs of Two Objects I—Velocity Equations

The graphs below show the velocity of two objects during the same time interval.

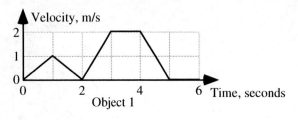

Object 1

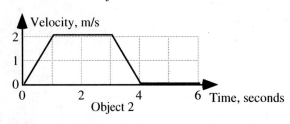

Object 2

Write the equations for the velocity in m/s as a function of time in seconds for these two motions for the first 5 seconds.

Time	Velocity Equation for Object 1	Velocity Equation for Object 2
0 s—1 s	$v(t) =$	$v(t) =$
1 s—2 s	$v(t) =$	$v(t) =$
2 s—3 s	$v(t) =$	$v(t) =$
3 s—4 s	$v(t) =$	$v(t) =$
4 s—5 s	$v(t) =$	$v(t) =$

NT3A-CCT19: BALL THROWN UPWARD AND COMES BACK DOWN—ACCELERATION

A ball is thrown straight upward and falls back to the same height. A student makes the graph of the speed of the ball as a function of time. Three students who are discussing this graph make the following contentions:

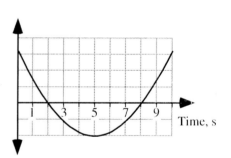

Akira: *"I don't think this can be correct because the sign of the acceleration changes on this graph, but the acceleration on the ball will be constant."*

Burt: *"No, I think this is right because it is only showing what happens to the speed, which will decrease to zero at the top and then increase as the ball falls. Since the slopes for both segments are the same except for sign that means the acceleration is constant."*

Catalina: *"This graph makes sense to me because it shows the speed decreasing. I disagree with Burt, because I think this means the acceleration is also decreasing until the ball gets to the top and stops. Then both the speed and acceleration increase as the ball falls down again."*

Which, if any, of these three students do you agree with and think is correct?

Akira _____ Burt _____ Catalina _____ None of them_____

Please explain your reasoning.

NT3A-QRT20: KINEMATICS GRAPHS—CHANGE DIRECTION

The graph at right is for an object in one-dimensional motion. The vertical axis is not labeled.

a) If the vertical axis is position, does the object ever change direction? If so, at what time or times does this change in direction occur?

Explain.

b) If the vertical axis is velocity, does the object ever change direction? If so, at what time or times does this change in direction occur?

Explain.

NT3A-QRT21: POSITION TIME GRAPHS OF TWO CHILDREN—KINEMATICS

The graph at right is of the motion of two children, Ariel and Byron, who are moving along a straight hallway. The vertical axis is not labeled.

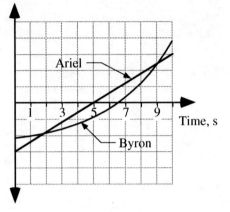

a) If the vertical axis is position, does either child ever change direction? If so, at what time or times does this change in direction occur?

Explain.

b) If the vertical axis is position, are the two children ever at the same position along the hallway? If so, at what time or times?

Explain.

c) If the vertical axis is position, do the two children ever have the same speed? If so, at what time or times?

Explain.

d) If the vertical axis is position, do the two children ever have the same acceleration? If so, at what time or times?

Explain.

e) If the vertical axis is velocity, do either of the children ever change direction? If so, at what time or times does this change in direction occur?

Explain.

f) If the vertical axis is velocity, do the two children ever have the same velocity? If so, at what time or times?

Explain.

g) If the vertical axis is velocity, do the two children ever have the same acceleration? If so, at what time or times?

Explain.

NT3B-WWT22: MOTION EQUATIONS IN ONE DIMENSION—AVERAGE VELOCITY

An object moves along the x-axis according to the following (with x in meters and t in seconds):

$$x = 10\text{m} - 4(\text{m/s})\,t + 2(\text{m/s}^3)\,t^3$$

A student is planning to calculate the object's average velocity between $t = 0$ and $t = 2$ seconds using the equation

$$v_{average} = \frac{v_{initial} + v_{final}}{2}$$

What, if anything, is wrong with this? If something is wrong, identify it and explain how to correct it. If this is correct, explain why.

NT3B-WBT23: POSITION EQUATION—PHYSICAL SITUATION

Describe the motion of an object that is represented by the equation below:

$$x = 33.6 \text{ m} - (2.8 \text{ m/s})t$$

NT3B-CCT24: Bicyclist on a Straight Road—Average Speed

Three students are discussing a situation where a bicyclist travels at a steady 18.0 m/s for 10 minutes, then at 6.0 m/s for 20 minutes and finally at 12.0 m/s for 15 minutes along a straight level road. Students make the following contentions about the bicyclist's average speed for the overall trip:

Aaron: *"I think the average speed for the entire period is 18 m/s because to find an average you sum the three values and divide by two."*

Bessie: *"I disagree. The average speed is 12 m/s because you add the three velocities, but then you have to divide by three."*

Cesar: *"No, you are both wrong. The average speed is 10.7 m/s because that is what you get when you divide 28,800 m, the total distance traveled on the straight road, by 2700 seconds, the total time it took."*

Which, if any, of these three students do you agree with?

Aaron_____ Bessie _____ Cesar _____ None of them_____

Please explain your reasoning.

NT3C-WWT25: Acceleration vs. Time Graph—Final Velocity

A student is given the acceleration versus time graph for a motorcyclist traveling along a straight level stretch of road. The student states:

"This motorcyclist was slowing down during the period up to 14 seconds because her acceleration was negative during this period."

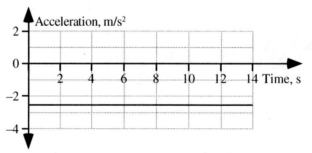

What, if anything, is wrong with this student's contention? If something is wrong, identify it, and explain how to correct it. If it is correct, explain why.

NT3C-QRT26: MOTION EQUATIONS IN ONE DIMENSION—LOCATION, VELOCITY, & ACCELERATION

An object moves along the *x*-axis according to this expression (with *x* in meters and *t* in seconds):

$$x = 10 - 4t + 2t^3$$

a) What is the object's location at *t* = 0?

b) What is the object's location at *t* = 2 sec?

c) What is the object's displacement between *t* = 0 and *t* = 2 sec?

d) What is the distance the object travels between *t* = 0 and *t* = 2 sec?

e) What is the object's velocity at *t* = 0?

f) What is the object's velocity at *t* = 2 sec?

g) What is the object's average velocity between *t* = 0 and *t* = 2 sec?

h) What is the object's acceleration at *t* = 0?

i) What is the object's acceleration at *t* = 2 sec?

j) What is the object's average acceleration between *t* = 0 and *t* = 2 sec?

NT3C-RT27: POSITION AS A FUNCTION OF TIME EQUATIONS—INSTANTANEOUS SPEED

The six equations below tell us the position in meters as a function of time in seconds for six objects that are moving along a straight line. As the equations show, these objects vary in their initial positions, initial velocities, and accelerations.

A. $x(t) = -7 + 9t - 2t^2$ **B.** $x(t) = +4 + 9t + t^2$ **C.** $x(t) = +3 - 7t - 2t^2$

D. $x(t) = -4 + 3t - 4t^2$ **E.** $x(t) = -1 - 9t - 2t^2$ **F.** $x(t) = -7 + t + 2t^2$

Rank these situations on the basis of the speed of the objects 2 seconds after the motions begin.

Greatest   1 _____ 2 _____ 3 _____ 4 _____ 5 _____ 6 _____ Least

OR, The speed at 2 seconds is the same but not zero for all these objects.   ____

OR, The speed at 2 seconds is zero for all these objects.   ____

OR, We cannot determine the ranking for the speeds of these objects.   ____

Please explain your reasoning.

NT3C-RT28: VELOCITY VS. TIME GRAPHS—INSTANTANEOUS VELOCITY

The graphs below show the velocity versus time for six boats traveling along a narrow channel that runs east to west. The scales on both axes are the same for all of these graphs, and east is positive. In each graph, a point is marked with a dot.

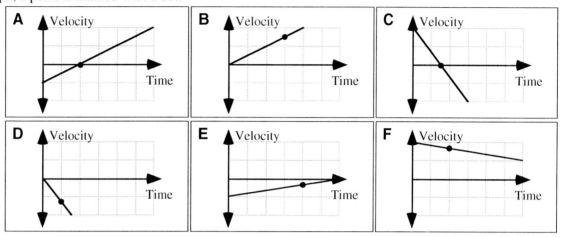

Rank these situations on the basis of the velocity of the boat at the point indicated.

Greatest   1 _____ 2 _____ 3 _____ 4 _____ 5 _____ 6 _____ Least

OR, The velocity at the marked points is the same but not zero for all these boats.   ____

OR, The velocity at the marked points is zero for all these boats.   ____

OR, We cannot determine the ranking for the velocity of these boats.   ____

Please explain your reasoning.

NT3C-RT29: VELOCITY VS. TIME GRAPHS—ACCELERATION

The graphs below show the velocity versus time for six boats traveling along a narrow channel that runs east to west. The scales on both axes are the same for all of these graphs, and east is positive. In each graph, a point is marked with a dot.

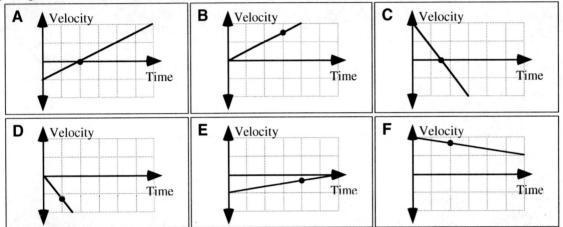

Rank these situations on the basis of the acceleration of the boat at the point indicated.

Greatest   1 _____ 2 _____ 3 _____ 4 _____ 5 _____ 6 _____ Least

OR, The acceleration at the marked points is the same but not zero for all these boats.   ____

OR, The acceleration at the marked points is zero for all these boats.   ____

OR, We cannot determine the ranking for the accelerations of these boats.   ____

Please explain your reasoning.

NT3C-RT30: Position vs. Time Graphs—Instantaneous Speed

The graphs below show the position versus time for six boats traveling along a narrow channel. The scales on both axes are the same for all of these graphs. In each graph, a point is marked with a dot.

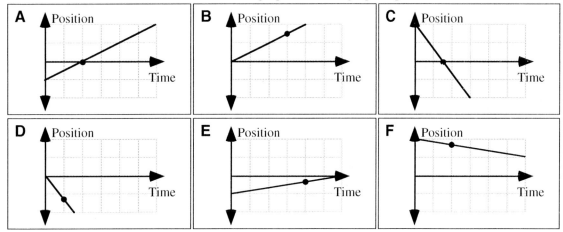

Rank these situations on the basis of the speed of the boat at the point indicated.

Greatest   1 _____ 2 _____ 3 _____ 4 _____ 5 _____ 6 _____ Least

OR, The speed at the marked points is the same for all these boats.   ____

OR, The speed at the marked points is zero for all these boats.   ____

OR, We cannot determine the ranking for the speeds of these boats.   ____

Please explain your reasoning.

NT3C-RT31: POSITION VS. TIME GRAPHS—DISPLACEMENT

The graphs below show the position versus time for six boats traveling along a narrow channel. The scales on both axes are the same for all of these graphs. In each graph, two points are marked with dots.

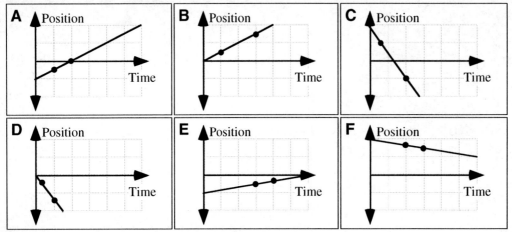

Rank these situations on the basis of the displacement between the two points.

Greatest   1 _____ 2 _____ 3 _____ 4 _____ 5 _____ 6 _____ Least

OR, The displacement between the points is the same but not zero for all the boats.   ____

OR, The displacement between the points is zero for all the boats.   ____

OR, We cannot determine the ranking for the displacements of these boats.   ____

Please explain your reasoning.

NT3C-RT32: POSITION VS. TIME GRAPHS—INSTANTANEOUS VELOCITY

The graphs below show the position versus time for six boats traveling along a narrow channel. The scales on both axes are the same for all of these graphs. In each graph, a point is marked with a dot.

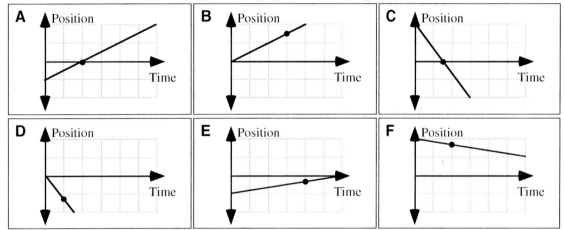

Rank these situations on the basis of the velocity of the boat at the point indicated.

Greatest   1 _____ 2 _____ 3 _____ 4 _____ 5 _____ 6 _____ Least

OR, The velocity at the marked points is the same but not zero for all the boats.   ___

OR, The velocity at the marked points is zero for all the boats.   ___

OR, We cannot determine the ranking for the velocities of these boats.   ___

Please explain your reasoning.

NT3C-CCT33: POSITION TIME EQUATION—INSTANTANEOUS SPEED

Three students are discussing the motion of an object which is described by the following equation:

$$x(t) = -4\text{m} - (9\text{m/s})t + (2\text{m/s}^2)t^2$$

The students make the following contentions about the motion during the first 2 seconds:

Amadeo: *"I think this object's instantaneous speed will increase with time since the acceleration is positive."*

Barrett: *"No, the object will have a decreasing speed, since the acceleration is directed opposite to the initial velocity."*

Chinue: *"I don't think we can tell what will happen to the speed from this equation, since it tells us about the position as a function of time, not about speed or velocity."*

Which, if any, of these three students do you agree with?

Amadeo_____ Barrett _____ Chinue _____ None of them_____

Please explain your reasoning.

NT3C-QRT34: POSITION TIME EQUATION—VELOCITY AND ACCELERATION EQUATION

A student is told that the motion of an object is described by the following equation:

$$x(t) = -4\text{m} - (9\text{m/s})t + (2\text{m/s}^2)t^2$$

What are the equations that describe the object's velocity and acceleration at various times?

$v(t) =$

$a(t) =$

Is the initial acceleration in the positive-*x* or negative-*x* direction?

Is the initial velocity in the positive-x or negative-x direction?

Is the initial position in the positive-x or negative-x direction from the origin?

After 3 seconds, is the acceleration in the positive-x or negative-x direction?

After 3 seconds, is the velocity in the positive-*x* or negative-*x* direction?

After 3 seconds, is the position in the positive-*x* or negative-*x* direction from the origin?

NT3E-QRT35: POSITION, VELOCITY, & ACCELERATION SIGNS—POSITION, DIRECTION, & RATE

Eight possible combinations for the signs for the instantaneous position, velocity, and acceleration of an object are given in the table below. Above the table is a coordinate axis that shows the origin, marked 0, and that indicates that the positive direction is to the right. The three columns on the right-hand side of the table are to describe the location of the object (either left or right of the origin), the direction of the velocity of the object (either toward or away from the origin), and what is happening to the speed of the object (either speeding up or slowing down at the given instant). The appropriate descriptions for the first case are shown.

Complete the rest of the table for position, direction, and rate.

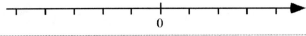

0

Position	Velocity	Acceleration	Position (Left or Right)	Direction (Toward or Away from)	Rate (Speeding up or Slowing down)
+	+	+	Right	Away from	Speeding up
+	+	−			
+	−	+			
+	−	−			
−	+	+			
−	+	−			
−	−	+			
−	−	−			

NT3E-WWT36: VELOCITY VS. TIME GRAPHS—ACCELERATION

The graphs show the velocity versus time for two boats traveling along a narrow channel that runs east to west. The scales on both axes are the same for the two graphs, and east is positive. In each graph, a point is marked with a dot. A student who is asked how the acceleration at the marked point for the object in graph A compares to the acceleration at the marked point for the object in graph B states:

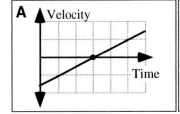
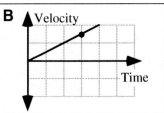

"I think that at the marked point the boat in graph B has the larger acceleration because the boat in graph A is at rest and its acceleration is zero."

What, if anything, is wrong with this student's contention? If something is wrong, identify it, and explain how to correct it. If it is correct, explain why.

NT3E-QRT37: Position, Velocity, & Acceleration Signs II—Position, Direction, & Rate

Eight possible signs combinations for the instantaneous position, velocity, and acceleration of an object are given in the table below. Above the table is a coordinate axis that shows the origin, marked 0, and that indicates that the positive direction is to the left. The three columns on the right-hand side of the table are to describe the location of the object (either left or right of the origin), the direction of the motion of the object (either toward or away from the origin), and what is happening to the speed of the object (either speeding up or slowing down at the given instant). The appropriate descriptions for the first case are shown.

Complete the table for the object's location relative to the origin, motion direction from the origin, and change in the speed. Please indicate any impossible combinations.

0

Position	Velocity	Acceleration	Position (Left or Right)	Direction (Toward or Away from)	Rate (Speeding up or Slowing down)
+	+	+	Left	Away from	Speeding up
+	+	−			
+	−	+			
+	−	−			
−	+	+			
−	+	−			
−	−	+			
−	−	−			

NT3F-WWT38: Bicyclist on a Hill—Velocity vs. Time Graph

A bicyclist moving at high speed on a straight road comes to a hill that slopes upward gradually. She decides to coast up the hill. A physics student observing the bicyclist plots the velocity-time graph for her trip up the hill as shown.

What, if anything, is wrong with this student's graph? If something is wrong, explain the error and how to correct it. If the graph is correct, explain why.

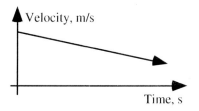

NT3F-RT39: Velocity vs. Time Graphs—Displacement of Identical Objects

Graphs of velocity versus time during 4 seconds for six identical objects are shown below. The objects move along a straight, horizontal surface under the action of a force exerted by an external agent.

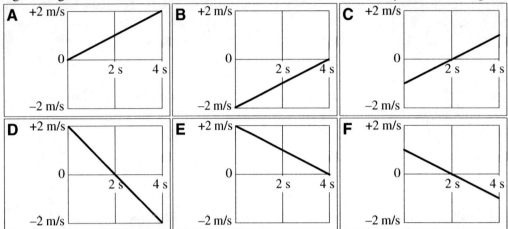

Rank these situations on the basis of the displacement of the objects during each of these intervals.

Greatest   1 _____ 2 _____ 3 _____ 4 _____ 5 _____ 6 _____ Least

OR, The displacement is the same for all these situations.   ____

OR, The displacement is zero for all these situations.   ____

OR, We cannot determine the ranking for the displacement for these situations. ____

Please explain your reasoning.

NT3F-RT40: Velocity vs. Time Graphs—Average Velocity of Identical Objects

Graphs of velocity versus time during 4 seconds for six identical objects are shown below. The objects move along a straight, horizontal surface under the action of a force exerted by an external agent.

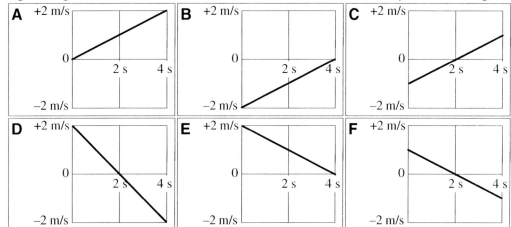

Rank these situations on the basis of the average velocity of the objects during each of these intervals.

Greatest   1 _____ 2 _____ 3 _____ 4 _____ 5 _____ 6 _____ Least

OR, The average velocity is the same but not zero for all these situations.   ____

OR, The average velocity is zero for all these situations.   ____

OR, We cannot determine the ranking of the average velocity for these situations.   ____

Please explain your reasoning.

NT3F-RT41: VELOCITY VS. TIME GRAPHS—ACCELERATION OF IDENTICAL OBJECTS

Graphs of velocity versus time during 4 seconds for six identical objects are shown below. The objects move along a straight, horizontal surface under the action of a force exerted by an external agent.

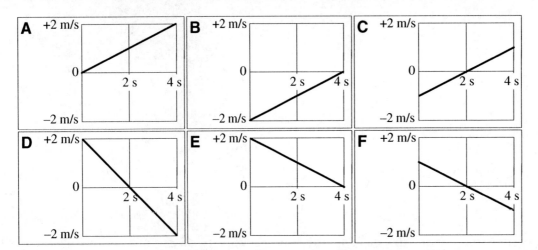

Rank these situations on the basis of the acceleration of these objects during each of these intervals.

Greatest   1 _____ 2 _____ 3 _____ 4 _____ 5 _____ 6 _____ Least

OR, The acceleration is the same but not zero for all these situations.   ____

OR, The acceleration is zero for all these situations.   ____

OR, We cannot determine the ranking of the acceleration for these situations.   ____

Please explain your reasoning.

NT3G-QRT42: POSITION VS. TIME GRAPHS—ACCELERATION AND VELOCITY

Position versus time graphs for boats traveling along a narrow channel are shown below. The scales on both axes are the same for all of these graphs. In each graph, a point is marked with a dot.

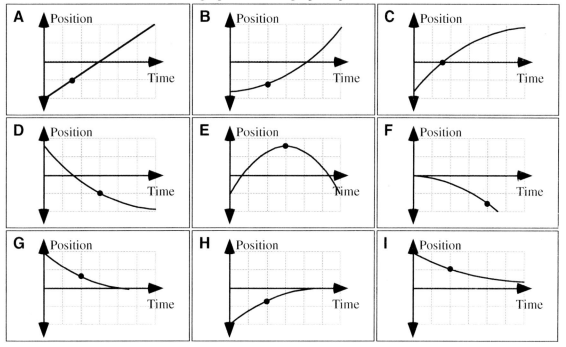

a) For which of these, if any, is the position zero at the indicated point?

b) For which of these, if any, is the position negative at the indicated point?

c) For which of these, if any, is the velocity zero at the indicated point?

d) For which of these, if any, is the velocity negative at the indicated point?

e) For which of these, if any, is the acceleration zero at the indicated point?

f) For which of these, if any, is the acceleration negative at the indicated point?

NT4 Motion in Two Dimensions

NT4A-CT1: Motorcycle Trips I—Displacement

Shown below are the paths two motorcyclists took on an afternoon ride. Both started at the same place. Rider A traveled east for 19 km and then south for 4 km. Rider B traveled south for 7 km and then east for 16 km.

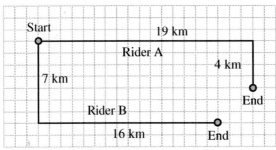

For these trips, is the magnitude of the displacement of Rider A *greater than*, *less than*, or *the same as* magnitude of the displacement of Rider B? Explain.

NT4A-CCT2: Motorcycle Trips II—Displacement

Shown below are the paths three motorcyclists took on an afternoon ride. Riders A and C traveled from the coffee shop to the mechanic's garage along different paths, while Rider B traveled from the garage to the coffee shop. Three physics students discussing these rides make the following contentions:

Ali: *"The lengths of the paths that Riders A and B travel are the same, so they have the same displacement. Rider C has the smallest displacement."*

Bob: *"I agree that Rider C has the smallest displacement, because the diagonal path is shortest. But the displacements of Riders A and B are actually different, because their directions are opposite each other."*

Carol: *"I think the displacements of all three riders are the same, because they go between the same two points. What path they follow doesn't matter."*

Which, if any, of these three students do you think is correct?

Ali _____ Bob _____ Carol _____ None of them_____

Please explain your reasoning.

NT4B-RT3: STUDENTS' JOURNEYS—AVERAGE VELOCITY

Six students went out for pizza to celebrate after completing their physics final. All six went directly from their residence hall to the nearby hangout, but they returned along the paths shown, taking different times. Values for the round-trip distances they traveled and the total times they took to walk their routes are given in the figures.

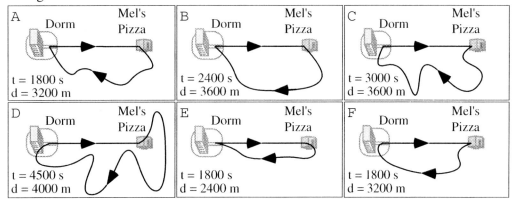

Rank these situations on the basis of the average velocity for the total trip.

Greatest   1 _____ 2 _____ 3 _____ 4 _____ 5 _____ 6 _____ Least

OR, The average velocity is the same but not zero for all these students' trips.   ____

OR, The average velocity is zero for all these students' trips.   ____

OR, We cannot determine the ranking for the average velocity here.   ____

Please explain your reasoning.

NT4B-WWT4: FALLING ROCK AND THROWN ROCK—VELOCITY GRAPHS

Rock *A* is dropped from the top of a cliff at the same instant that Rock *B* is thrown horizontally away from the cliff. The rocks are identical. A student draws the following graphs to describe part of the motion of the rocks. He uses a coordinate system in which up is the positive vertical direction, and the positive horizontal direction is away from the cliff, with the origin at the point the rocks were released.

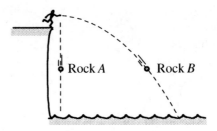

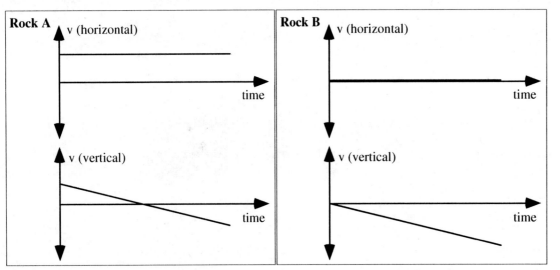

What, if anything, is wrong with these graphs for the motions of the two rocks? If something is wrong, identify it and explain how to correct it. If the graphs are correct, explain why.

NT4B-CCT5: Car on a Country Road—Average Speed and Velocity

An automobile travels along a winding country road, speeding up and slowing down as it goes. The car ends up 30 km directly north of its starting point after 40 minutes of travel. Three students make the following contentions about this situation:

Arlo: *"I think if we calculated the average velocity and average speed for the car, the average velocity would have a larger value. The car moved in two dimensions, so the velocity, which is a vector, will be greater than the speed, which is a scalar quantity."*

Bunmi: *"I don't see any reason the average velocity and average speed would have different values. They are just different names for the same thing."*

Cici: *"I think the average speed will have a larger value. The road is not straight, so the distance traveled will be more than the displacement."*

Which, if any, of these three students do you agree with and think is correct?

Arlo _____ Bunmi _____ Cici _____ None of them_____

Please explain your reasoning.

NT4B-QRT6: Velocity & Position of the Moon—Velocity Change Direction

The drawing at right shows the position of the moon at two times, about seven days apart. The velocity vector for the moon at each of these times is shown.

Find the direction of the change in velocity of the moon in this time interval. If the change in velocity is zero, state that explicitly.

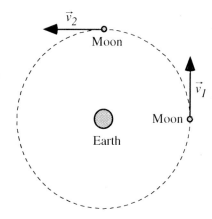

NT4C-CCT7: MOTORCYCLE ON ROAD COURSE—ACCELERATION

A motorcycle is slowing down as it travels through a bend in a road. The path of the motorcycle is the dashed line shown in the bird's-eye view. The arrow represents the motorcycle's velocity at the instant shown. Three physics students make the following contentions about the acceleration of the motorcycle:

Alexi: *"The motorcycle's acceleration is in the opposite direction to the velocity since it is slowing down."*

Bindi: *"No, the acceleration will have two components, one opposite the velocity and the other toward the center of the curve."*

Carlos: *"I don't think the motorcycle has an acceleration, since it is braking."*

Which, if any, of these three students do you think is correct?

Alexi _____ Bindi _____ Carlos _____ None of them _____

Please explain your reasoning.

NT4C-CCT8: TRUCK SLOWING DOWN—AVERAGE ACCELERATION DIRECTION

A truck is originally driving north. It slows down as it turns a corner, as shown in the top-view drawing. At point B, the truck is driving west. Three students are discussing the direction of the average acceleration of the truck as it moves from point A to point B.

Alisa: *"From the velocity vectors we can find the change in velocity by subtracting the velocity vector at A from the velocity vector at B. The change in velocity vector points in the same direction as the acceleration, which will be a bit south of southwest."*

Bernie: *"The truck is slowing down, so the acceleration direction is toward the rear of the truck. The acceleration vector changes continuously as the truck goes around the corner. At A, it is pointing south, and by B, it is pointing east. It never has a westward component — that's just crazy talk."*

Collette: *"Between A and B, the truck is moving in a circle, so the acceleration is centripetal, or toward the center of the circle. If we think of the road between A and B as part of a circle, then the center is exactly southwest."*

Which, if any, of these students do you agree with?

Alisa _____ Bernie _____ Colette _____ None of them _____

Please explain your reasoning.

NT4C-CT9: ICE GLIDERS ON A FROZEN LAKE—AVERAGE ACCELERATION

Two ice gliders (essentially large sleds with sails) are sliding on a frozen lake. Glider A goes from traveling northwest at 30 m/s to traveling northwest at 20 m/s in 3 seconds while glider B goes from traveling northwest at 10 m/s to traveling southeast at 10 m/s in 6 seconds.

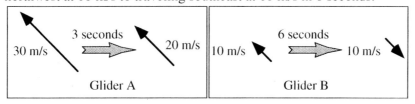

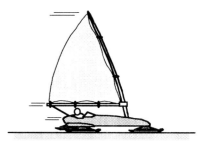

For these intervals, will the magnitude of the average acceleration of Glider A be *greater than, less than,* **or** *equal to* **the magnitude of the average acceleration of Glider B?**

Explain.

NT4C-WWT10: SPEEDBOATS CHANGING VELOCITIES—ACCELERATION

Two speedboats are racing on a lake. Boat A goes from traveling at 15 m/s east to 20 m/s north in 10 seconds. Boat B goes from 10 m/s west to 15 m/s west in 10 seconds.

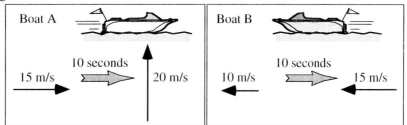

A student watching the race states:

> *"These two boats have the same acceleration for the 10-second interval since they both changed their velocities by 5 m/s in that time interval."*

What, if anything, is wrong with this student's contention? If something is wrong, identify it, and explain how to correct it. If the contention is correct, explain why.

NT4C-QRT11: MOTION EQUATIONS IN TWO DIMENSION—LOCATION, VELOCITY, & ACCELERATION

An object moves in the xy plane with its position given by the following equations with x and y in meters and t in seconds:

$$x = 6m + (4m/s^2)t^2$$
$$y = 10m - (2m/s^2)t^2$$

a. What is the object's location at $t = 0$?

b. What is the object's location at $t = 1$ sec?

c. What is the object's displacement between $t = 0$ and $t = 2$ sec?

d. What is the object's velocity at $t = 0$?

e. What is the object's velocity at $t = 2$ sec?

f. What is the object's average velocity between $t = 0$ and $t = 2$ sec?

g. What is the object's acceleration at $t = 0$?

h. What is the object's acceleration at $t = 2$ sec?

i. What is the object's average acceleration between $t = 0$ and $t = 2$ sec?

nT4D-QRT12: Constant Speed Car on Oval Track—Acceleration & Velocity Directions

A car on an oval track travels at a constant speed and moves clockwise around the track.

In the table below, indicate the direction of the velocity and acceleration of the car for the labeled points. Use the direction labels in the rosette at far right, **J** for no direction, **K** for into the page, **L** for out of the page, or **M** if none of these are correct.

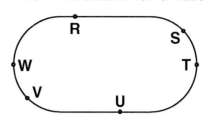 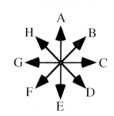

Point on track	Velocity direction	Acceleration direction
R		
S		
T		
U		
V		
W		

nT4D-QRT13: Car on Oval Track—Direction of the Acceleration and Velocity

A car on an oval track starts from rest at point R and moves clockwise around the track. It increases its speed at a constant rate until it reaches point T, and then travels at a constant speed until it returns to point R.

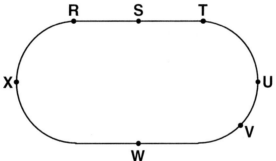 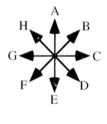

At the points given in the table, indicate the direction of the velocity and acceleration of the car. Use the direction labels in the rosette to the right of the racetrack drawing, **J** for no direction, **K** for into page, **L** for out of page, or **M** if none of these are correct.

Point on track	Velocity direction	Acceleration direction
S		
U		
V		
W		
X		

NT4D-QRT14: Car on Track—Direction of the Acceleration and Velocity

A car on a track starts from rest at point R and moves clockwise around the track shown in the bird's-eye view below. It increases its speed at a constant rate until it reaches point T, and then travels at a constant speed until it reaches point W. From point W to point Y, the speed of the car decreases at a constant rate, and it comes to rest at point Y.

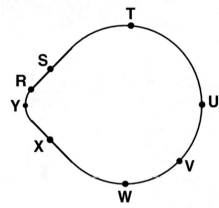

At the points given in the table, indicate the direction of the velocity and acceleration of the car. Use the direction labels in the rosette to the right of the drawing of the track, **J** for no direction, **K** for into page, **L** for out of page, or **M** if none of these are correct.

Point on track	Velocity direction	Acceleration direction
S		
U		
V		
X		

NT4E-CCT15: Rifle Shots—Time to Hit Ground

Rifles are fired horizontally from platforms at various heights. The bullets fired from these rifles are identical, but they leave the rifle barrels at different speeds as shown in the diagrams. All of the bullets miss their targets and hit the ground. Ignore air resistance in this task.

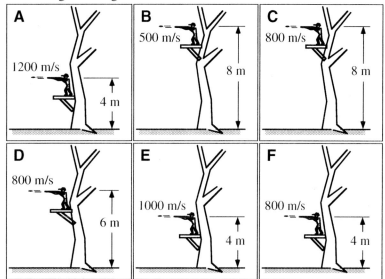

Students who are asked to rank these situations on the basis of how long it takes the bullets to hit the ground respond as follows:

Anja: *"I think the ranking should be C > B > D > A> E > F, because if two bullets are shot from the same height at different speeds, their y-acceleration is the same, meaning the one shot faster would have to cover more of the horizontal distance before hitting the ground, thereby making the time longer. So we rank first by height then by velocity."*

Brina: *"The higher the platform, the longer it will take, but the faster the bullet the smaller the time to hit the ground. So using rate times time equals distance we get time = height/velocity, which gives us the ranking B > C > D > F > E> A."*

Charlie: *"I think the ranking should be A > E > C > D > F > B. I agree that the height of the platform matters as does the velocity. The faster a bullet is moving, the longer it takes to hit the ground and the higher the longer too. So we rank first by velocity, then by the height if the velocities are the same."*

Deepa: *"I get B = C > D > A = E = F. The time that each bullet is in the air depends on the initial vertical velocity and the height. Since the initial vertical velocity is zero we only need to worry about the height, with the larger height giving a longer time. The horizontal velocity does not matter."*

Ellie: *"I think the ranking is A > E > C = D = F > B, since the time to reach the ground is directly related to the horizontal velocity."*

Which, if any, of these students do you agree with?

Anja ___ Brina ___ Charlie ___ Deepa ___ Ellie___ None of them___

Explain.

NT4E-QRT16: PROJECTILE MOTION—VELOCITY AND ACCELERATION GRAPHS

A baseball is thrown from point *S* in right field to home plate. The dashed line in the diagram shows the path of the ball. For this exercise, use a coordinate system with up as the positive vertical direction and to the right as the positive horizontal direction, with the origin at the point the ball was thrown from (point *S*). Ignore air resistance.

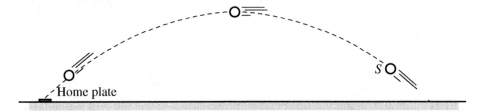

Sketch below and explain graphs for:

(1) The horizontal velocity vs. time and the vertical velocity vs. time.

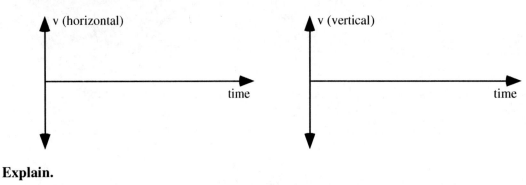

Explain.

(2) The horizontal acceleration vs. time and the vertical acceleration vs. time.

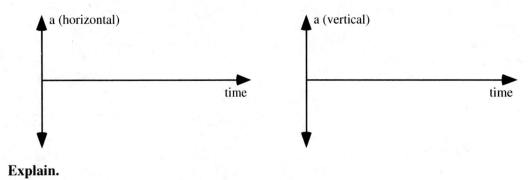

Explain.

nT4E-WWT17: PROJECTILE MOTION—VELOCITY AND ACCELERATION GRAPHS

A student draws the following graphs for the velocity and acceleration of the motion of a rock that is thrown upward and toward the west. She uses a coordinate system in which up is the positive vertical direction and toward the west is the positive horizontal direction, with the origin at the point the rock was released. For this exercise you should ignore air resistance.

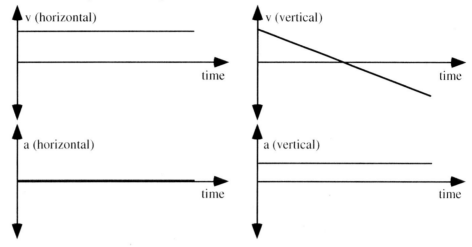

What, if anything, is wrong with the above graphs? If something is wrong, explain the error and how to correct it. If the graphs are correct, explain why.

NT4E-QRT18: Projectile Motion—Velocity and Acceleration Graphs for Two Rocks

Two rocks are either dropped or thrown at the same time from the same cliff. Graphs for part of the motion of the rocks are shown for a coordinate system in which up is the positive vertical direction and the positive horizontal direction is away from the cliff. The origin is at the point at which the rocks were released. For this exercise, you should ignore air resistance.

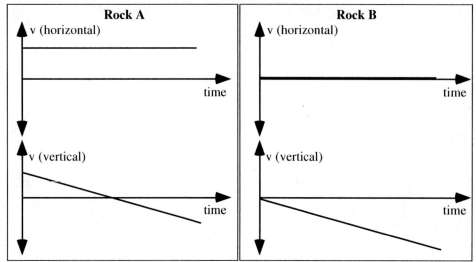

(1) Is Rock *A dropped, thrown,* or *is it not possible to determine whether it was dropped or thrown?* Explain.

(2) Is Rock *B dropped, thrown,* or *is it not possible to determine whether it was dropped or thrown?* Explain.

(3) Does Rock *A* hit the ground *first, at the same time,* or *after* Rock *B?* Explain.

(4) Does Rock *A* hit the ground *closer to, at the same distance from,* or *farther from* the base of the cliff compared to Rock *B?* Explain.

NT4E-QRT19: PROJECTILE MOTION FOR TWO ROCKS—VELOCITY AND ACCELERATION GRAPHS I

Two identical rocks are thrown horizontally from a cliff with Rock A having a greater velocity at the instant it is released than Rock B. For this exercise, you should ignore air resistance. Use a coordinate system with up as the positive vertical direction, away from the cliff as the positive horizontal direction, and with the origin at the point of release at the top of the cliff.

a) Sketch velocity vs. time graphs for each of the rocks.

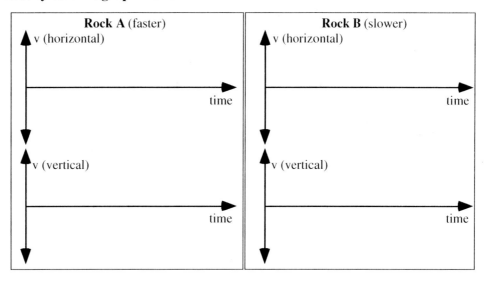

b) Which rock hits the ground first?

c) Which rock lands farthest from the base of the cliff?

Explain why you drew the graphs as you did and how you determined your answers.

NT4E-QRT20: PROJECTILE MOTION FOR TWO ROCKS—VELOCITY AND ACCELERATION GRAPHS II

Two identical rocks are thrown horizontally from a cliff with Rock A having a greater velocity at the instant it is released than Rock B. For this exercise, you should ignore air resistance. Use a coordinate system with down as the positive vertical direction, away from the cliff as the positive horizontal direction, and with the origin at the bottom of the cliff directly below the release point,

a) Sketch below the velocity vs. time graphs for each of the rocks.

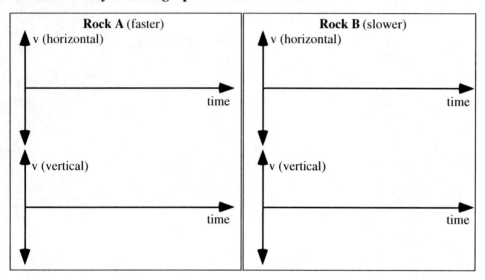

b) Which rock hits the ground first?

c) Which rock lands farthest from the base of the cliff?

Explain why you drew the graphs as you did and how you determined your answers.

NT4E-QRT21: Baseball Projectile Motion—Velocity and Acceleration Graphs

A baseball is thrown from point S in right field to home plate. The dashed line in the diagram shows the path of the ball.

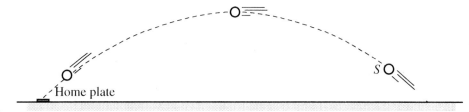

For this exercise, use a coordinate system with up as the positive vertical direction and to the left as the positive horizontal direction, and with the origin at home plate.

Select the graph from the choices below that best represents:

1) the horizontal velocity vs. time graph _____

2) the horizontal acceleration vs. time graph _____

3) the vertical velocity vs. time graph _____

4) the vertical acceleration vs. time graph _____

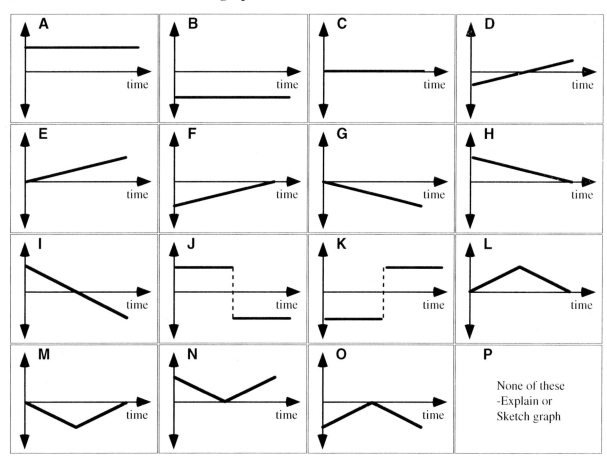

nT4E-CRT22: Projectile Motion for Two Rocks—Velocity and Acceleration Graphs II

Two students throw two rocks horizontally from a cliff with different velocities. Both rocks hit the water below at the same time but Rock *B* hits farther from the base of the cliff. For this exercise, you should ignore air resistance. Use coordinates where up is the positive vertical direction, away from the cliff is the positive horizontal direction, and the origin is at the top of the cliff at the point of release.

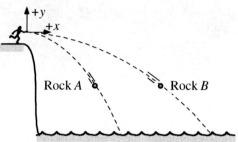

Sketch below velocity vs. time graphs for each rock.

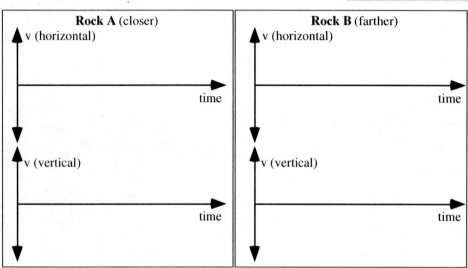

Explain how you determined your answer.

NT4E-LMCT23: DROPPED PRACTICE BOMB—LANDING POINT

A military airplane is flying 1200 m above the ground at a speed of 200 m/s. It drops a practice bomb that hits the ground after traveling a horizontal distance of 3130 m. Ignore air resistance.

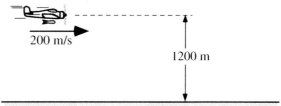

For each of the numbered changes below, use the lettered choices below to identify what will happen to the horizontal distance the bomb travels while falling compared to the situation above.

(a) The horizontal distance will be *greater than* 3130 m.
(b) The horizontal distance will be *less than* 3130 m but not to zero.
(c) The horizontal distance will be *equal to* 3130 m.
(d) The horizontal distance will be *zero*, i.e., the bomb will drop straight down.
(e) We *cannot determine* how this change will affect the horizontal distance.

For each of the following changes, only the feature(s) identified is(are) modified from the given situation above.

1) The plane's speed is tripled. _____
  Explain.

2) The plane is climbing straight up at the release point. _____
  Explain.

3) The plane is flying in level flight at an altitude of 1100 m. _____
  Explain.

4) The mass of the bomb is increased. _____
  Explain.

5) The bomb is thrown from the plane with a vertical downward velocity of 15 m/s. _____
  Explain.

6) The plane is diving at a 20° angle and is at a height of 1200 m. _____
  Explain.

7) The plane's speed decreases, and it is flying at an altitude of 1800 m. _____
  Explain.

NT4E-LMCT24: CANNONBALL FIRED FROM AN ELEVATED CANNON—MAXIMUM HEIGHT

A cannonball is fired from a cannon located at the edge of a cliff. The muzzle of the cannon is 40 meters above the water. The cannonball has an initial horizontal velocity of 80 meters per second and an initial upward vertical muzzle velocity of 45 meters per second.

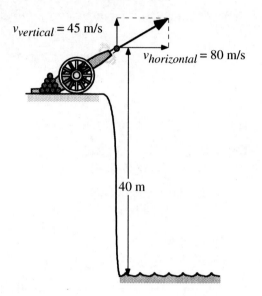

$v_{vertical}$ = 45 m/s

$v_{horizontal}$ = 80 m/s

40 m

Identify, from choices (a) – (d), how each change described below will affect the maximum height that the cannonball reaches above the water level.

This change will:
  (a) *increase* the maximum height of the cannonball.
  (b) *decrease* the maximum height of the cannonball.
  (c) *have no effect* on the maximum height of the cannonball.
  (d) have an effect on the maximum height of the cannonball that *cannot be determined.*

All of these modifications are changes to the initial situation that is shown in the drawing.

1) **The cannon is adjusted so that the initial vertical velocity of the projectile is increased and the initial horizontal velocity is unchanged.**

   Explain.

2) **The gun is adjusted so only the initial horizontal velocity of 150 m/s of the projectile is reduced and the initial vertical velocity is unchanged.**

   Explain.

3) **The cannon is moved to a lower cliff.**

   Explain.

4) **The mass of the cannonball is reduced without changing the initial velocities.**

   Explain.

NT4E-CT25: Toy Trucks Rolling Off Tables with Different Heights—Time

Two toy trucks roll off the ends of tables. The heights of the tables, the speeds of the trucks, and the masses of the trucks are given.

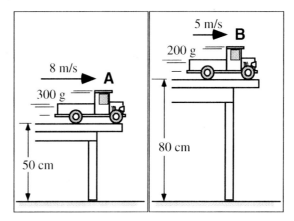

Will Truck A be in the air for *a longer, a shorter*, or *the same* time as Truck B before it reaches the floor?

Explain.

NT4E-CT26: Toy Trucks with Different Speeds Rolling Off Identical Tables—Time

Two toy trucks roll off the ends of identical tables. The speeds and masses of the trucks are given.

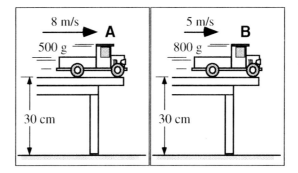

Will Truck A be in the air for *a longer, a shorter*, or *the same* time as Truck B before it reaches the floor?

Explain.

NT4E-CT27: PROJECTILE MOTION FOR TWO ROCKS—VELOCITY AND ACCELERATION

Two identical rocks are thrown horizontally from a cliff with different velocities. The rocks are thrown at the same time, and are shown after a few seconds. Neglect air resistance.

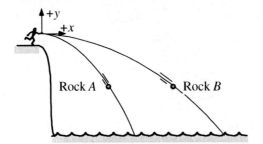

For the instant shown:

1) Will the magnitude of the horizontal velocity of Rock *A* be *greater than, less than,* or *equal to* the magnitude of the horizontal velocity of Rock *B*?

   Explain.

2) Will the magnitude of the vertical velocity of Rock *A* be *greater than, less than,* or *equal to* the magnitude of the vertical velocity of Rock *B*?

   Explain.

3) Will the magnitude of the horizontal acceleration of Rock *A* be *greater than, less than,* or *equal to* the magnitude of the horizontal acceleration of Rock *B*?

   Explain.

4) Will the magnitude of the vertical acceleration of Rock *A* be *greater than, less than,* or *equal to* the magnitude of the vertical acceleration of Rock *B*?

   Explain.

NT4E-CRT28: PROJECTILE MOTION FOR TWO ROCKS—VELOCITY AND ACCELERATION

Two identical rocks are thrown horizontally from a 60-meter tall cliff with initial velocities of 20 m/s and 30 m/s. The rocks are thrown at the same time, and are shown after a few seconds. Neglect air resistance. Use the coordinate system with the origin at the place where the rocks are released and with directions as shown in the diagram.

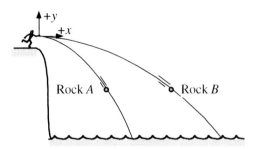

Write the position equations for Rock A, x(t) and y(t), and the velocity equations for Rock B, $v_x(t)$ and $v_y(t)$.

Rock A:

$x(t) =$

$y(t) =$

Explain.

Rock B:

$v_x(t) =$

$v_y(t) =$

Explain.

NT5A-RT1: MOVING SPACESHIP WITH FOUR CARGO PODS—TENSION IN RODS

Shown is a spaceship pulling four cargo pods at a constant velocity. The pods are connected to each other by rods, and a rod connects Pod A to the spaceship. The velocity of the spaceship and of the pods is 5000 m/s. All masses are given in the diagram in terms of *M*, the mass of an empty pod. (Since this is in space, we can ignore any resistive forces.)

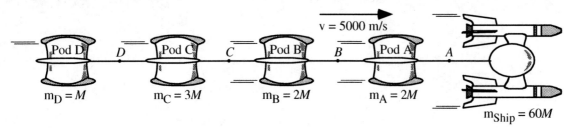

Rank the magnitude of the tension at the labeled points in the rods.

Greatest   1 _____ 2 _____ 3 _____ 4 _____   Least

OR, The magnitude of the tension in all the rods is the same but not zero.   ____

OR, The magnitude of the tension in all the rods is zero.   ____

OR, The ranking for the tensions in the rods cannot be determined.   ____

Explain your reasoning.

NT5A-RT2: CARTS ON INCLINES—MAGNITUDE OF THE NET FORCE

In each of the six figures below, a cart that has a motor and brakes is traveling either up or down an incline at a constant speed. The carts are identical but they carry either a 2 kg or 4 kg load and are on one of two inclines. Incline angles, cart masses, and speeds are given in each figure.

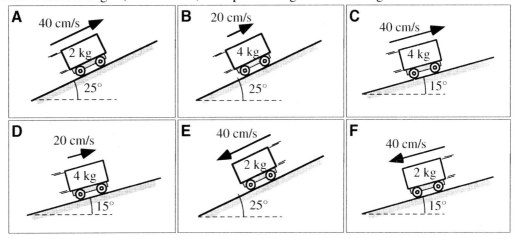

Rank these situations on the basis of the magnitude of the net force acting on the cart.

Greatest   1 _____ 2 _____ 3 _____ 4 _____ 5 _____ 6 _____ Least

OR, The net force is the same but not zero for all the carts.   ____

OR, The net force is zero for the carts.   ____

OR, We cannot determine the ranking for the net force on these carts.   ____

Please explain your reasoning.

NT5A-RT3: Moving Spaceships with Two Cargo Pods—Tension in Rods

In both cases shown, a moving spaceship is connected to two cargo pods, one empty and one full. In each case, a rod connects the spaceship to Pod 1 and another rod connects Pod 1 to Pod 2. In Case A, the speed of the ship and of the pods is 300 m/s, while in Case B it is 500 m/s. Both speeds are constant. All masses are given in the diagram in terms of M, the mass of an empty pod. (Since this is in space, we can ignore any resistive forces.)

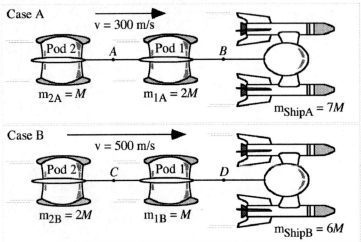

Rank the magnitude of the tensions at the labeled points in rods.

> Greatest 1 _____ 2 _____ 3 _____ 4 _____ Least

OR, The magnitude of the tension in all the rods is the same but not zero. ____

OR, The magnitude of the tension in all the rods is zero. ____

OR, The ranking for the tensions in the rods cannot be determined. ____

Explain your reasoning.

NT5B-RT4: CURLER PUSHING STONE—FORCE ON STONE

The figures below show six identical curling stones (the playing pieces in the sport of curling) that are being pushed horizontally along the ice by the thrower. For each stone, the instantaneous velocity and acceleration of the stones are given. The positive direction is to the right. Assume the ice is frictionless for the curling stones.

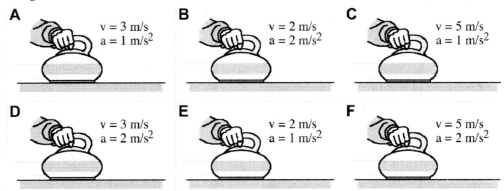

A v = 3 m/s, a = 1 m/s²

B v = 2 m/s, a = 2 m/s²

C v = 5 m/s, a = 1 m/s²

D v = 3 m/s, a = 2 m/s²

E v = 2 m/s, a = 1 m/s²

F v = 5 m/s, a = 2 m/s²

Rank these stones on the basis of the magnitude of the force the thrower is exerting on them at the instant shown.

Greatest   1 _____ 2 _____ 3 _____ 4 _____ 5 _____ 6 _____ Least

OR, The magnitude of the force by the thrower is the same for all the stones but not zero. ____

OR, The magnitude of the force by the thrower is zero for all the stones. ____

OR, We cannot determine the ranking for the magnitude of the forces. ____

Please explain your reasoning.

NT5B-RT5: Tugboat Pushing Barges—Force Tugboat Exerts on Lead Barge

In each of the six figures below, a tugboat is pushing two barges. The systems of tugboats and barges are accelerating at different rates. As shown in the figures, the barges have different loads so they have different masses, and the tugboats have different masses also. The accelerations of the systems are given for each case. The positive direction is to the right. Ignore the effects of fluid friction.

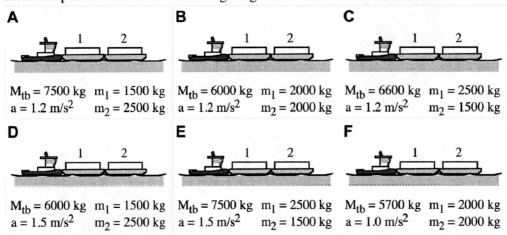

A

$M_{tb} = 7500$ kg $m_1 = 1500$ kg
$a = 1.2$ m/s^2 $m_2 = 2500$ kg

B

$M_{tb} = 6000$ kg $m_1 = 2000$ kg
$a = 1.2$ m/s^2 $m_2 = 2000$ kg

C

$M_{tb} = 6600$ kg $m_1 = 2500$ kg
$a = 1.2$ m/s^2 $m_2 = 1500$ kg

D

$M_{tb} = 6000$ kg $m_1 = 1500$ kg
$a = 1.5$ m/s^2 $m_2 = 2500$ kg

E

$M_{tb} = 7500$ kg $m_1 = 2500$ kg
$a = 1.5$ m/s^2 $m_2 = 1500$ kg

F

$M_{tb} = 5700$ kg $m_1 = 2000$ kg
$a = 1.0$ m/s^2 $m_2 = 2000$ kg

Rank these six cases on the basis of the magnitude of the force the tugboat exerts on barge two.

Greatest   1 _____ 2 _____ 3 _____ 4 _____ 5 _____ 6 _____ Least

OR, The force the tugboats exert on barge two is the same but not zero for all the systems.   ____

OR, The tugboats do not exert a force on barge two in any of the systems.   ____

OR, We cannot determine the ranking for the forces the tugboats exert on barge two.   ____

Please explain your reasoning.

NT5B-RT6: TUGBOAT PUSHING BARGES—FORCE TUGBOAT EXERTS ON FIRST BARGE

Each of the six figures below shows a system consisting of a tugboat pushing two barges labeled 1 and 2. The masses of the tugboats and the barges along with the accelerations of the systems are given for each case. The positive direction is to the right. Ignore the effects of fluid friction.

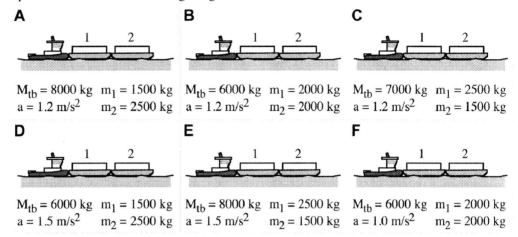

A

M_{tb} = 8000 kg m_1 = 1500 kg
a = 1.2 m/s² m_2 = 2500 kg

B

M_{tb} = 6000 kg m_1 = 2000 kg
a = 1.2 m/s² m_2 = 2000 kg

C

M_{tb} = 7000 kg m_1 = 2500 kg
a = 1.2 m/s² m_2 = 1500 kg

D

M_{tb} = 6000 kg m_1 = 1500 kg
a = 1.5 m/s² m_2 = 2500 kg

E

M_{tb} = 8000 kg m_1 = 2500 kg
a = 1.5 m/s² m_2 = 1500 kg

F

M_{tb} = 6000 kg m_1 = 2000 kg
a = 1.0 m/s² m_2 = 2000 kg

Rank these six cases on the basis of the magnitude of the force the tugboat exerts on barge one.

Greatest   1 _____ 2 _____ 3 _____ 4 _____ 5 _____ 6 _____ Least

OR, The force the tugboats exert on barge one is the same but not zero for all the systems. ____

OR, The tugboat does not exert a force on barge one for any of these systems. ____

OR, We cannot determine the ranking for the force the tugboat exerts on barge one. ____

Please explain your reasoning.

NT5B-CT7: CURLER PUSHING STONE—FORCE ON STONE

The figures below show two identical curling stones (the playing pieces in the sport of curling) that are being pushed horizontally by the thrower. The instantaneous speed and acceleration of the two stones are given. The positive direction is to the right. Ignore the friction between the stone and the ice.

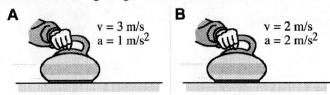

A   v = 3 m/s   a = 1 m/s^2

B   v = 2 m/s   a = 2 m/s^2

Is the magnitude of the force that the thrower is exerting on the stone in Case A *greater than, less than,* or *equal to* the magnitude of the force that the thrower is exerting in situation B?

Explain.

NT5B-WBT8: THREE EQUATIONS—PHYSICAL SITUATION

The equations below result from the application of Newton's Laws to a system of three blocks that may be on rough surfaces:

$$T_2 - (0.16)(0.84 \text{ kg})(9.8 \text{ m/s}^2) = (0.84 \text{ kg})a$$

$$T_1 - T_2 - (0.33)(0.45 \text{ kg})(9.8 \text{ m/s}^2)(\cos 26°) - (0.45 \text{ kg})(9.8 \text{ m/s}^2)(\sin 26°) = (0.45 \text{ kg})a$$

$$(6.65 \text{ kg})(9.8 \text{ m/s}^2) - T_1 = (6.65 \text{ kg})a$$

Draw a physical situation that would result in these equations and explain how your drawing is consistent with these equations.

NT5B-RT9: TIME-VARYING FORCE ON A CART—ACCELERATION

The graph below shows the net force in the *x*-direction acting on a cart as a function of time.

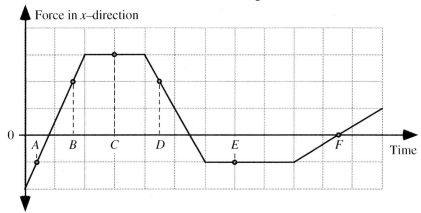

Rank the acceleration of the cart at the labeled times.

Greatest Positive 1 _____ 2 _____ 3 _____ 4 _____ 5 _____ 6 _____ Greatest Negative

OR, The acceleration of the cart is the same but not zero for all these times.   ____

OR, The acceleration of the cart is zero for all these times.   ____

OR, The ranking of the acceleration of the cart cannot be determined.   ____

Explain your reasoning.

NT5B-RT10: BLOCKS FLOATING IN LIQUIDS—FORCE EXERTED BY LIQUID

The six figures below show identical blocks that have been placed into six different liquids. There is the same volume of liquid in all six containers. All of the blocks are at rest. The masses of the liquids vary and are specified in the figures.

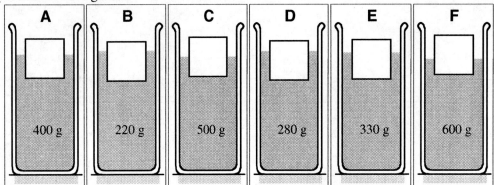

Rank these six liquids on the basis of the force they exert on the block.

Greatest   1 _____ 2 _____ 3 _____ 4 _____ 5 _____ 6 _____ Least

OR, The force exerted by the liquids is the same for all these liquids.   ___

OR, The liquids exert zero force in all these cases.   ___

OR, We cannot determine the ranking for these forces.   ___

Please explain your reasoning.

NT5B-RT11: Mass on a Stretched Horizontal Spring System—Initial Acceleration

The figures below show systems containing a block initially held at rest on a frictionless surface. In each system, the block is attached to the end of a spring, which is stretched to the right. The mass and spring constant are given for each system, as well as the distance the spring is initially stretched. When the mass is released, the spring will accelerate the block.

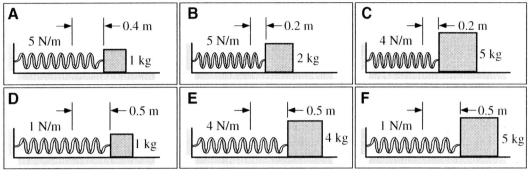

Rank these systems on the basis of the magnitude of the initial acceleration of the blocks.

Greatest 1 _____ 2 _____ 3 _____ 4 _____ 5 _____ 6 _____ Least

OR, The magnitude of the initial accelerations of the blocks will be the same for all these blocks. ___

OR, We cannot determine the ranking for the initial accelerations of the blocks. ___

Please explain your reasoning.

NT5B-QRT12: STUDENT PUSHING TWO BLOCKS—FORCE

A student pushes horizontally on two blocks. The blocks are moving to the right. Block *A* has more mass than block *B*. There is friction between the blocks and the table.

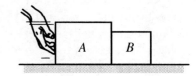

(1) For the situation where the blocks are moving at a constant speed, which of the following statements is true about the magnitude of the forces?

a) The force that block *A* exerts on block *B* is *greater than* the force that block *B* exerts on block *A*.
b) The force that block *A* exerts on block *B* is *less than* the force that block *B* exerts on block *A*.
c) The force that block *A* exerts on block *B* is *equal to* the force that block *B* exerts on block *A*.
d) We cannot compare the forces unless we know how fast the blocks are slowing down.

Explain.

(2) For the situation where the blocks are moving at a constant speed, which of the following statements is true about the net force?

a) The net force on block *A* points *to the right* and is *equal to* the net force on block *B*.
b) The net force on block *A* points *to the left* and is *equal to* the net force on block *B*.
c) The net force on block *A* points *to the right* and is *greater than* the net force on block *B*.
d) The net force on block *A* points *to the left* and is *greater than* the net force on block *B*.
e) The net force on block *A* points *to the right* and is *less than* the net force on block *B*.
f) The net force on block *A* points *to the left* and is *less than* the net force on block *B*.
g) None of these are correct.

Explain.

(3) For the situation where the blocks are slowing down, which of the following statements is true about the magnitude of the forces?

a) The force that block *A* exerts on block *B* is *greater than* the force that block *B* exerts on block *A*.
b) The force that block *A* exerts on block *B* is *less than* the force that block *B* exerts on block *A*.
c) The force that block *A* exerts on block *B* is *equal to* the force that block *B* exerts on block *A*.
d) We cannot compare the forces unless we know how fast the blocks are slowing down.

Explain.

(4) For the situation where the blocks are slowing down, which of the following statements is true about the net force?

a) The net force on block *A* points *to the right* and is *equal to* the net force on block *B*.
b) The net force on block *A* points *to the left* and is *equal to* the net force on block *B*.
c) The net force on block *A* points *to the right* and is *greater than* the net force on block *B*.
d) The net force on block *A* points *to the left* and is *greater than* the net force on block *B*.
e) The net force on block *A* points *to the right* and is *less than* the net force on block *B*.
f) The net force on block *A* points *to the left* and is *less than* the net force on block *B*.
g) None of these are correct.

Explain.

nT5B-RT13: Ropes Pulling Identical Boxes—Rope Tension

Shown below are boxes that are being pulled by ropes along frictionless surfaces, accelerating toward the left. All of the boxes are identical, and the accelerations of the boxes are indicated in each figure.

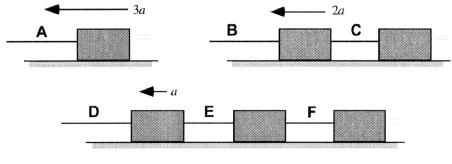

Rank the magnitude of the tension in these ropes.

Largest 1_____ 2_____ 3_____ 4_____ 5_____ 6_____ Smallest

OR, All of these ropes have the same tension. ____

OR, We cannot determine the ranking for the tensions. ____

Please explain your reasoning.

nT5B-RT14: ROPES PULLING IDENTICAL BOXES—ROPE TENSION

Shown below are boxes that are being pulled by ropes along frictionless surfaces, accelerating toward the left. All of the boxes are identical, and the accelerations of the boxes are indicated in each figure.

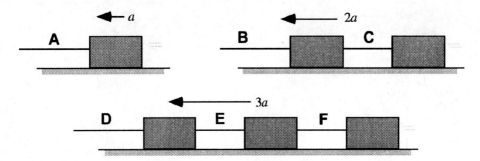

Rank the magnitude of the tension in these ropes.

      Largest 1_____ 2_____ 3_____ 4_____ 5_____ 6_____ Smallest

OR, All of these ropes have the same tension.   ___

OR, We cannot determine the ranking for the tensions.   ___

Please explain your reasoning.

NT5B-QRT15: SKATEBOARD RIDER COASTING DOWN A HILL—ACCELERATION AND NET FORCE

At the instant shown, a skateboard rider is coasting down a hill and speeding up. Consider the skateboard and the rider as a single system and ignore friction.

a) **Use velocity vectors to find the approximate direction of the acceleration of the system.**

b) **Draw a free-body diagram for the system labeling all forces, and show how the forces in your free-body diagram add to give a net force in the direction of the acceleration.**

nT5B-RT16: ROPES PULLING BOXES—ROPE TENSION

Shown below are boxes that are being pulled by ropes along frictionless surfaces, accelerating toward the left. The accelerations of the boxes are the same, and the masses of the boxes are indicated in each figure.

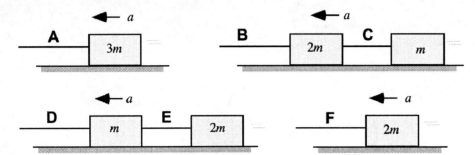

Rank the magnitude of the tension in these ropes.

Largest 1_____ 2_____ 3_____ 4_____ 5_____ 6_____ Smallest

OR, All of these ropes have the same tension. ____

OR, We cannot determine the ranking for the tensions. ____

Please explain your reasoning.

nT5B-CT17: Ropes Pulling Boxes—Rope Tension

Shown below are boxes that are being pulled by a force F along frictionless surfaces, accelerating toward the left. The masses of the boxes are indicated in each figure.

Will the magnitude of the tension in rope A on the left be *greater than, less than,* **or** *equal to* **the magnitude of the tension in rope B on the right?**

Explain.

NT5B-WWT18: Lifting Up a Pail—Strategy

A loaded pail is attached to a rope that passes around an overhead pulley and is tied to a ring on the floor. Linda, a construction worker, plans to untie the rope from the ring, pull on the rope to lift the pail 1 meter higher, and then retie the rope. Linda weighs 800 N and is capable of lifting twice her weight, 1600 N. The loaded pail weighs 1200 N.

What, if anything, is wrong with Linda's plan? Explain how to correct it or if the plan will work, explain why.

NT5B-CCT19: BLOCK MOVING AT CONSTANT SPEED—FORCES ON BLOCK

A student uses a string to pull a block across a table at a constant speed of 2 meters per second. The string makes an angle θ with the horizontal. A second student makes a free-body diagram of the block, and then uses this free-body diagram to generate a vector sum diagram as shown.

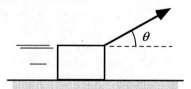

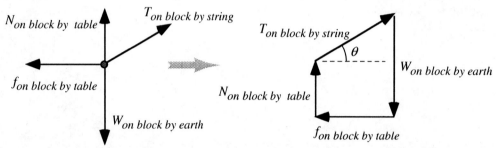

Three students are comparing the magnitudes of the forces in the vector sum diagram:

Anja: *The vector sum diagram allows us to compare the magnitudes of all four forces: The weight is the largest, then the tension, then friction, then the normal force.*

Barb: *Well, the weight is definitely greater than the normal force. But there should be a net force to the right in the vector sum because that's the way the block is moving, and there isn't. I don't think we can use it to rank the other forces.*

Cole: *I think we can use it to say that the weight is greater than the normal force. Also, the tension is greater than the friction, since the friction is the same length as the dashed line, and this is equal to the tension times the cosine of theta (θ). But we can't compare the vertical forces with the horizontal ones.*

Which, if any, of these students do you agree with?

Anja _____ Barb _____ Cole _____ None of them _____

Please explain your reasoning.

NT5B-WWT20: VELOCITY VS. TIME GRAPHS—NET FORCE

These graphs below show the velocity versus time for two identical train engines on a straight track. A positive velocity indicates that the engine was traveling east. The scales on both axes are the same for the graphs. On each graph a point is marked with a dot.

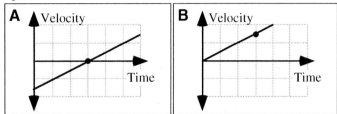

A student who is asked how the net force acting on the engine in graph A at the identified point compares to the net force acting on the engine in graph B states:

"I think that B has the larger net force since the net force on A at the identified point is zero."

What, if anything, is wrong with this statement? If something is wrong, identify it, and explain how to correct it. If the statement is correct, explain why.

NT5B-CT21: BLOCKS MOVING AT CONSTANT SPEED—FORCE ON BLOCK

In both cases shown, a block is moving to the right across a rough table with a constant speed of 2 m/s. The tables and the blocks are identical. In Case A, the block is pushed with a stick and in Case B, the block is pulled with a string. The angle that the applied force makes with the horizontal is the same in both cases.

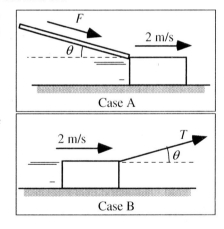

Will the magnitude of the force on the block by the stick in Case A be *greater than, less than,* or *equal to* the tension on the block by the string in Case B?

Explain.

NT5B-CRT22: OBJECT MOVING IN ONE DIMENSION—FORCE VS. TIME

An object moves along a straight path from its
starting point to a second point, and then returns to
its starting point. The graph of the speed of the
object as a function of time is shown to the right.

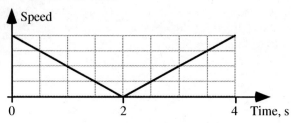

**On the axes below, draw the force versus time
graph for this motion.**

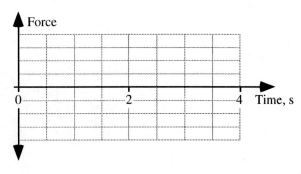

Explain.

NT5B-WWT23: PULLING A BLOCK ACROSS A ROUGH SURFACE—FORCE RELATIONSHIPS

A person pulls a block across a rough horizontal surface at a
constant speed by applying a force **F** at a slight angle as shown.
A free-body diagram is drawn for the block. The arrows in the
diagram correctly indicate the directions, but not necessarily the
magnitudes of the various forces on the block. A student makes
the following claim about this free-body diagram:

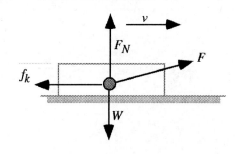

*"The velocity of the block is constant, so the net force acting on
the block must be zero. Thus the normal force F_N equals the
weight W, and the force of friction f_k equals the applied force F."*

**What, if anything, is wrong with this statement? If something is wrong, identify it and explain how
to correct it. If this statement is correct, explain why.**

NT5B-CT24: Masses on a Pulley—Supporting Force

In both cases shown, a pulley is suspended by a rope from a ceiling, and two masses are tied to each other by a rope that passes around the pulley. In Case A the two masses are at rest, but in Case B the masses are accelerating. The pulleys are identical in the two cases, and all ropes are massless.

Will the tension at point *P* in the rope between the pulley and the ceiling be *greater in Case A, greater in Case B*, or *the same in both cases*?

Explain.

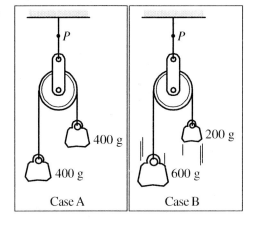

NT5B-CRT25: Velocity vs. Time—Force vs. Time

Shown is the velocity versus time graph for an object that is moving in one dimension under the (perhaps intermittent) action of a single horizontal force.

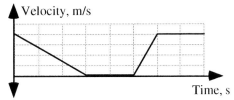

On the axes below draw the horizontal force acting on this object as a function of time.

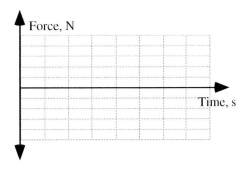

Explain.

NT5B-CT26: Spaceships Pulling Two Cargo Pods—Tension in Tow Rods

In both cases shown below a spaceship is pulling two cargo pods, one empty and one full. At the instant shown, the speed of the pods and spaceships is 300 m/s. In Case A the acceleration of the ship and of the pods is 3 m/s^2, while in Case B it is 2 m/s^2. All masses are given in terms of M, the mass of an empty pod.

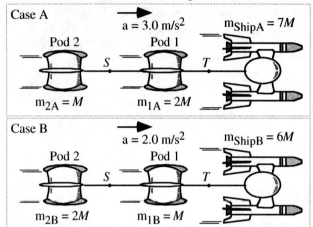

a) Will the tension at point S in the tow rod be *greater in Case A*, *greater in Case B*, or *equal in both cases*?

Explain.

b) Will the tension at point T in the tow rod be *greater in Case A*, *greater in Case B*, or *equal in both cases*?

Explain.

NT5B-QRT27: THREE VECTORS—RESULTANT

a) In the space below, add the three vectors shown to the right and label the resultant vector as $\vec{R}$.

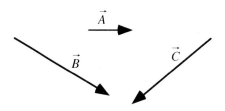

Suppose the three vectors above represent forces exerted on a slice of pepperoni pizza by three people, Abel ($\vec{A}$), Beth ($\vec{B}$), and Celia ($\vec{C}$) as shown in the top view picture to the right. A fourth person, David, also pulls on the pizza. The pizza moves to the left at a constant speed. Assume there is no friction between the pizza slice and the greasy table.

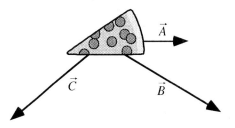

b) In what direction is David pulling on the pizza?

Explain your reasoning.

NT5B-RT28: Moving Spaceship Pulling Four Cargo Pods—Tension in Tow Rods

A spaceship is pulling four cargo pods. At the instant shown, the velocity of the spaceship and of the pods is 1000 m/s, and they have an acceleration of 3 m/s^2 in the same direction as the velocity. All masses are given in the diagram in terms of M, the mass of an empty pod.

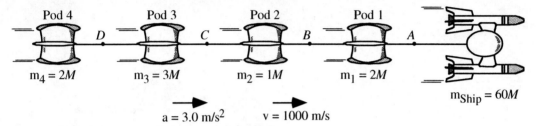

Rank the magnitude of the tension in the tow rods at the labeled points.

Greatest   1 _____ 2 _____ 3 _____ 4 _____   Least

OR, The magnitude of the tension in all the tow rods is the same but not zero.   ___

OR, The magnitude of the tension in all the tow rods is zero.   ___

OR, The ranking for the tensions in the tow rods cannot be determined.   ___

Explain your reasoning.

NT5B-CT29: SPACESHIPS PULLING TWO CARGO PODS—TENSION OR COMPRESSION IN TOW RODS

In each case below a spaceship is attached to two cargo pods by rods. At the instant shown the speed of the pods and of the spaceship is 300 m/s. In Case A the acceleration of the ship and of the pods is 3 m/s^2 to the left, while in Case B it is 2 m/s^2 to the right. All masses are given in terms of M, the mass of an empty pod.

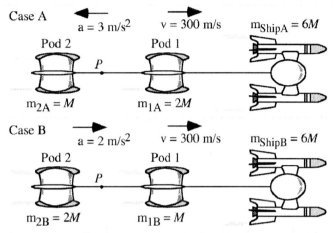

Case A

$a = 3$ m/s^2 $v = 300$ m/s $m_{ShipA} = 6M$

Pod 2   Pod 1

P

$m_{2A} = M$ $m_{1A} = 2M$

Case B

$a = 2$ m/s^2 $v = 300$ m/s $m_{ShipB} = 6M$

Pod 2   Pod 1

P

$m_{2B} = 2M$ $m_{1B} = M$

Will the magnitude of the tension or compression at point P in the tow rod be *greater in Case A*, *greater in Case B*, or *the same in both cases*?

Explain.

NT5B-RT30: STACKED BLOCKS SPEEDING UP ON A CONVEYOR BELT—NET FORCE

Various stacks of blocks are traveling along a conveyer belt. At the instant shown, all blocks have the same velocity of 3 m/s to the right, and the same acceleration of 2 m/s², also to the right. The blocks do not slip. All masses are given in the diagram in terms of M, the mass of the smallest block.

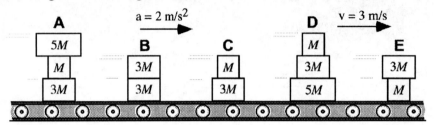

Rank the magnitude of the net force on each stack of blocks.

Greatest   1 _____ 2 _____ 3 _____ 4 _____ 5 _____   Least

OR, The magnitude of the net force on all the stacks is the same but not zero.   ____

OR, The magnitude of the net force on all the stacks is zero.   ____

OR, The ranking for the net forces for the stacks cannot be determined.   ____

Explain your reasoning.

NT5B-RT31: Spaceships Moving Two Cargo Pods—Tension or Compression in Rods

A spaceship is attached by tow rods to two cargo pods. At the instant shown, the speed of the pods and of the spaceship is 300 m/s. In Case A the acceleration of the ship and of the pods is 3 m/s² to the left, while in Case B it is 3 m/s² to the right. All masses are given in terms of M, the mass of an empty pod.

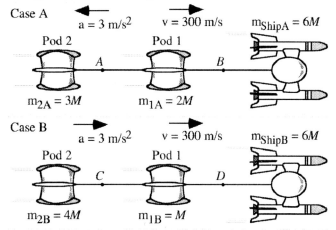

Rank the magnitude of the tension or compression at the labeled points in the tow rods.

Greatest   1 _____ 2 _____ 3 _____ 4 _____   Least

OR, The magnitude of the tension or compression in all the rods is the same but not zero.   ___

OR, The magnitude of the tension or compression in all the tow rods is zero.   ___

OR, The ranking for the tension or compression in the tow rods cannot be determined.   ___

Explain your reasoning.

NT5C-QRT32: THROWN BASEBALL—FREE-BODY DIAGRAM AT THE TOP

A baseball is thrown from right field to home plate (HP), traveling from right to left in the diagram.

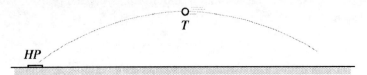

A group of physics students watching the game create the following free-body diagrams for the baseball at the top of its path (point *T*). Note that the forces are not drawn to scale.

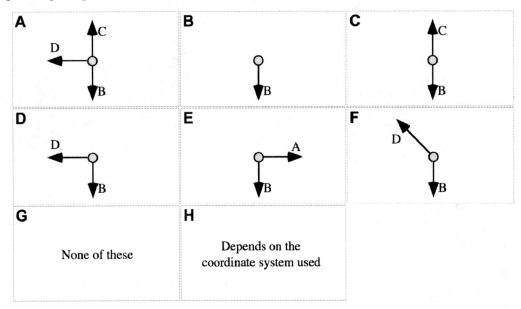

1a) If they decide to *ignore air friction,* which is the correct free-body diagram for the baseball at point *T*?

1b) Define all forces on the ball for this force diagram:

2a) If they decide to *include air friction,* which is the correct free-body diagram for the baseball at point *T*?

2b) Define all forces on the ball for this force diagram:

NT5C-WWT33: BOX ON INCLINE—FORCES

A heavy box is sitting at rest on an incline. There is friction between the box and the incline and a rope is pulling on the box in a direction up and to the left, parallel to the incline. A physics student draws a free-body diagram below for the box.

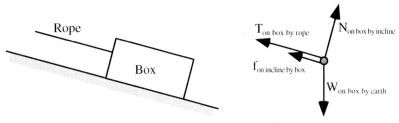

What, if anything, is wrong with this student's free-body diagram? If something is wrong, explain the error and how to correct it. If this free-body diagram is correct, explain why.

NT5C-QRT34: SUITCASE SPEEDING UP AS IT SLIDES DOWN RAMP—FORCES ON SUITCASE

A suitcase is speeding up as it slides down a ramp angled at 45° to the horizontal.

Draw a free-body diagram labeling all the forces on the suitcase, and then rank the magnitudes of the forces you have drawn.

Explain your ranking.

NT5C-WWT35: Thrown Baseball—Free-body Diagram for Ascending Baseball

A baseball is thrown from right field to home plate (HP), traveling from right to left in the diagram.

A physics student watching the game produces the free-body diagram shown below for the baseball as it moves upward at a point *(S)* along the path. She explains that in her drawing: "I'm ignoring air resistance. **W** is the weight of the baseball and **D** is the force of the throw on the baseball."

What, if anything, is wrong with this free-body diagram? If something is wrong, identify it and explain how to correct it. If this free-body diagram is correct, explain why.

NT5C-QRT36: Suitcase Sliding Down Ramp at Constant Speed—Forces on Suitcase

A suitcase is moving at a constant speed as it slides down a ramp angled at 45° to the horizontal.

Draw a free-body diagram labeling all the forces on the suitcase, and then rank the magnitudes of the forces you have drawn.

Explain your ranking.

NT5C-TT37: BOX ON BOX ON TABLE—FREE-BODY DIAGRAMS

Box *A* is on a table with box *B* on top of it. A horizontal rope is pulling box *A* to the right with a force *F* as shown. There is friction between the boxes and between the lower box and the surface it is traveling along. The top box *B* is at rest relative to the bottom box *A*, and both boxes are accelerating to the right.

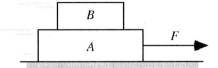

A student draws separate (not-to-scale) free-body diagrams for boxes *A* and *B* and defines all the forces as shown below.

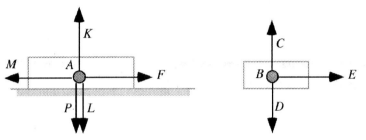

K: Normal force on *A* by the table
L: Weight or gravitational force on *A* by the earth
M: Kinetic frictional force on *A* by the table
F: Tension force on *A* by the rope
P: Normal force on *A* by *B*

D: Weight or gravitational force on *B* by the earth
E: Static frictional force on *B* by *A*
C: Normal force on *B* by *A*

There is at least one thing wrong with these free-body diagrams. Please identify all errors(s) and explain how to correct the diagram(s).

NT5C-TT38: THROWN BASEBALL—FORCE DIAGRAM

A baseball is thrown from right field to home plate (HP), traveling from right to left in the diagram.

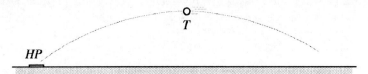

A student watching the game sketches the force(s) on the baseball at the top of its path (point *T*). He *ignores air friction* and notes that the forces are not drawn to scale in his sketch. He produces the following diagram for the forces on the ball at the top:

The student states:

> *"A is the horizontal component of the initial applied force, B is the vertical component of the initial applied force, and C is the force of gravity."*

There is a problem with the student's description. Explain what is wrong, and correct it.

NT5C-WWT39: BOX ON INCLINE—FORCE DIAGRAM

A heavy box is sitting at rest on an incline. There is friction between the box and the incline and a rope is pulling on the box in a direction up and to the left, parallel to the incline. A student sketches an unscaled free-body diagram for the box.

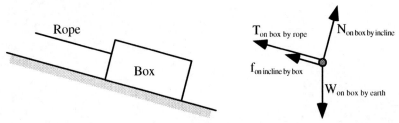

What, if anything, is wrong with the student's diagram? If something is wrong, identify it, and explain how to correct it. If the student's diagram is correct, explain why.

NT5C-RT40: Box Sliding on Moving Box on Table—Horizontal Forces in the Free-Body Diagrams

Box *A* is on a table with box *B* on top of it. A horizontal rope is pulling box *A* to the right with a force *F* as shown. There is friction between the boxes, and between the lower box and the table. The top box *B* is sliding relative to the bottom box while box *A* is accelerating to the right. For all pairs of surfaces, the coefficient of static friction is 0.6 and the coefficient of kinetic friction is 0.4. The weight of box *B* is half of the weight of box *A*.

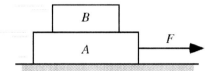

A student draws separate free-body diagrams for boxes *A* and *B* and defines all the forces as shown below. (The forces are not drawn to scale on these diagrams.)

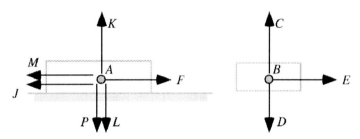

J: Kinetic frictional force on *A* by *B*
K: Normal force on *A* by the table
L: Weight or gravitational force on *A* by the earth
M: Kinetic frictional force on *A* by the table
F: Tension force on *A* by the rope
P: Normal force on *A* by *B*

D: Weight or gravitational force on *B* by the earth
E: Kinetic frictional force on *B* by *A*
C: Normal force on *B* by *A*

Rank the magnitudes of the horizontal forces in the free-body diagrams above.

Greatest   1 _____ 2 _____ 3 _____ 4 _____   Least

OR, The magnitudes of the horizontal forces are the same but not zero.   ____

OR, The ranking for the magnitudes of the horizontal forces cannot be determined.   ____

Explain your reasoning.

NT5D-CT41: IDENTICAL TOY TRUCK COLLISIONS—FORCE AND ACCELERATION

Shown below are two identical toy trucks traveling at different constant speeds that are about to collide.

1) The trucks are traveling in the same direction.

Will the magnitude of the force exerted on truck A by truck B be *greater than, less than,* or *equal to* the magnitude of the force exerted on truck B by truck A?

Explain.

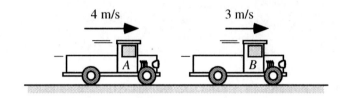

2) The trucks are traveling in opposite directions.

Will the magnitude of the force exerted on truck A by truck B be *greater than, less than,* or *equal to* the magnitude of the force exerted on truck B by truck A?

Explain.

3) The trucks are traveling in the same direction.

Will the magnitude of the acceleration of truck A during the collision be *greater than, less than,* or *equal to* the magnitude of the acceleration of truck B during the collision?

Explain.

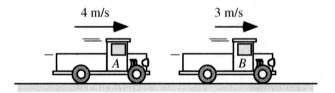

4) The trucks are traveling in opposite directions.

Will the magnitude of the acceleration of truck A during the collision be *greater than, less than,* or *equal to* the magnitude of the acceleration of truck B during the collision?

Explain.

NT5D-CT42: TOY TRUCK COLLISIONS—FORCE ON TRUCKS

Shown below are situations where two toy trucks traveling at different constant speeds are about to collide.

1) The two identical trucks are traveling in the same direction, and truck *B* is carrying a heavy load.

Will the magnitude of the force exerted on truck *A* by truck *B* be *greater than, less than,* or *equal to* the magnitude of the force exerted on truck *B* by truck *A*?

Explain.

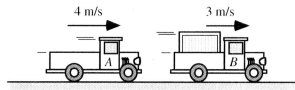

2) The two identical trucks are traveling in opposite directions, and truck *B* is carrying a heavy load.

Will the magnitude of the force exerted on truck *A* by truck *B* be *greater than, less than,* or *equal to* the magnitude of the force exerted on truck *B* by truck *A*?

Explain.

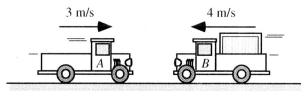

3) The two identical trucks are traveling in the same direction, and truck *A* is carrying a heavy load.

Will the magnitude of the force exerted on truck *A* by truck *B* be *greater than, less than,* or *equal to* the magnitude of the force exerted on truck *B* by truck *A*?

Explain.

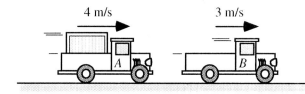

4) The two identical trucks are traveling in opposite directions, and truck *A* is carrying a heavy load.

Will the magnitude of the force exerted on truck *A* by truck *B* be *greater than, less than,* or *equal to* the magnitude of the force exerted on truck *B* by truck *A*?

Explain.

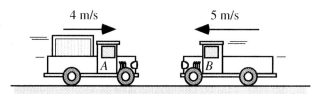

NT5D-CT43: Toy Truck Collisions—Acceleration

Shown below are situations where two toy trucks traveling at different constant speeds are about to collide.

1) The two identical trucks are traveling in the same direction, and truck *B* is carrying a heavy load.

Will the magnitude of the acceleration of truck *A* during the collision be *greater than, less than,* or *equal to* the magnitude of the acceleration of truck *B* during the collision?

Explain.

2) The two identical trucks are traveling in opposite directions, and truck *B* is carrying a heavy load.

Will the magnitude of the acceleration of truck *A* during the collision be *greater than, less than,* or *equal to* the magnitude of the acceleration of truck *B* during the collision?
Explain.

3) The two identical trucks are traveling in the same direction, and truck *A* is carrying a heavy load.

Will the magnitude of the acceleration of truck A during the collision be *greater than, less than*, or *equal to* the magnitude of the acceleration of truck B during the collision?

Explain.

NT5D-RT44: STACKED BLOCKS—FORCE DIFFERENCES

Shown are stacks of various blocks that are sitting at rest. All masses are given in the diagram in terms of M, the mass of the smallest block. The labels $A - G$ in the diagram refer to surfaces where the blocks are in contact with one another.

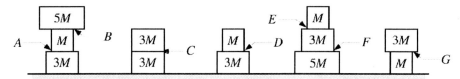

Rank the magnitudes of force *differences* between the forces the blocks exert on each other at the labeled surfaces. (For example, at surface D it would be the difference between the force exerted on the smaller upper block by the lower block and the force exerted on the larger lower block by the upper block.)

Greatest   1 _____ 2 _____ 3 _____ 4 _____ 5 _____ 6 _____ 7 _____   Least

OR, The magnitude of the force differences at the surfaces is the same but not zero. ____

OR, The magnitude of the force differences at the surfaces is zero. ____

OR, The ranking for the force differences at the surfaces cannot be determined. ____

Explain your reasoning.

NT5D-WWT45: TENNIS BALL AND RACQUET—FORCE

A tennis player returns a serve. A physics student watching the match makes the following contention:

"While the ball is in contact with the racquet, the racquet exerts a larger force on the ball than the ball does on the racquet because the racquet has to stop the ball and then reverse its motion."

What, if anything, is wrong with this contention? If something is wrong, explain the error and how to correct it. If this contention is correct, explain why.

NT5D-WWT46: BALL HITTING A WALL—FORCES

A student observes a rubber ball hitting a wall and rebounding. She states:

"In this situation the wall exerts a larger force on the ball than the ball exerts on the wall, because the ball undergoes an acceleration but the wall doesn't move. That is, the ball goes from an initial speed to zero, and then from zero to the rebound speed, but the wall does not accelerate since it is stationary the whole time."

What, if anything, is wrong with this contention? If something is wrong, identify it, and explain how to correct it. If this contention is correct, explain why.

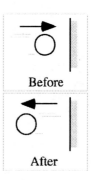

Before

After

NT5D-WWT47: ROCK ON TABLE—FORCES AND REACTION FORCES

A rock with a weight of 10 N is resting on a table. A student makes a number of statements about this situation.

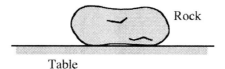

1) *"The weight of the rock is a force of 10 N by gravity in the downward direction."*

What, if anything, is wrong with this statement? If something is wrong, identify it and explain how to correct it. If this statement is correct, explain why.

2) *"The reaction force to this weight is a force of 10 N exerted on the rock by the table in the upward direction."*

What, if anything, is wrong with this statement? If something is wrong, identify it and explain how to correct it. If this statement is correct, explain why.

3) *"The normal force exerted on the rock by the table is a force of 10 N; the reaction force to this normal force is a force of 10 N exerted on the rock by gravity in the downward direction."*

What, if anything, is wrong with this statement? If something is wrong, identify it and explain how to correct it. If this statement is correct, explain why.

4) *"If the 10 N rock is lifted off the table by a hand that exerts a force of 12 N upward on the rock, the reaction force to this 12 N force is a force of 10 N exerted on the hand by the rock in the downward direction."*

What, if anything, is wrong with this statement? If something is wrong, identify it and explain how to correct it. If this statement is correct, explain why.

NT5D-LMCT48: IDENTICAL TOY TRUCKS COLLIDING HEAD-ON—FORCE ON TRUCKS

Two identical toy trucks traveling at the same speed in opposite directions are about to collide. The magnitude of the force exerted on truck *A* by truck *B* during the collision is equal to the magnitude of the force exerted on truck *B* by truck *A*.

4 m/s   4 m/s

Identify from choices (a)-(d) how each change described below will affect the magnitude of the force exerted on truck *A* by truck *B* during the collision as compared to the magnitude of the force exerted on truck *B* by truck *A*.

This change will cause the magnitude of the force exerted on truck *A* by truck *B* to be:

(a) *greater than* the magnitude of the force exerted on truck *B* by truck *A*.

(b) *less than* the magnitude of the force exerted on truck *B* by truck *A*.

(c) *the same* as the magnitude of the force exerted on truck *B* by truck *A*.

(d) *indeterminate as compared to* the magnitude of the force exerted on truck *B* by truck *A*.

All of these modifications are changes to the initial situation shown in the diagram.

1) Truck *A* is carrying a heavy load.   _____

    Explain.

2) Truck *A* is going faster than truck *B*.   _____

    Explain.

3) Truck *A* is carrying a heavy load and is going faster than truck *B*.   _____

    Explain.

4) Truck *A* is speeding up (accelerating).   _____

    Explain.

5) Truck *A* is speeding up (accelerating) and is carrying a heavier load.   _____

    Explain.

6) Truck *A* is speeding up (accelerating) while truck *B* is carrying a heavier load.   _____

    Explain.

NT5E-CT49: Person in an Elevator Moving Upward—Scale Reading

In the two cases shown below, a person is standing on a scale in an elevator. The elevators are identical, and the person weighs 500 N. In both cases the elevator is moving upward, but in Case A it is accelerating upward and in Case B it is accelerating downward.

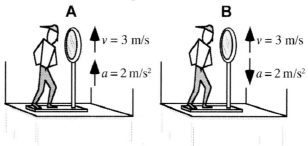

Will the scale reading in Case A be *greater than, less than,* or *the same as* the scale reading in Case B?

Explain.

NT5E-CT50: Person in an Elevator Moving Downward—Scale Reading

In the two cases shown below, a person is standing on a scale in an elevator. The elevators are identical, and the person weighs 500 N. In both cases the elevator is moving downward, but in Case A it is accelerating upward and in Case B it is accelerating downward.

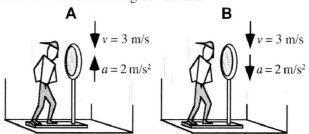

Will the scale reading in Case A be *greater than, less than,* or *the same as* the scale reading in Case B?

Explain.

NT5E-CT51: Person in an Elevator—Scale Reading

In the two cases shown below, a person is standing on a scale in an elevator. The elevators are identical, and the person weighs 500 N. In Case A the elevator is moving upward at a constant speed, and in Case B the elevator is moving downward at a constant speed.

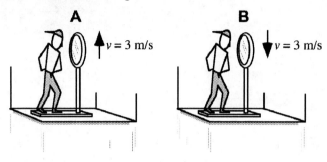

Will the scale reading in Case A be *greater than*, *less than*, or *the same as* the scale reading in Case B?

Explain.

NT5E-CT52: Skateboarder on Circular Bump—Weight and Normal Force

A skateboarder is skating over a circular bump. At the instant shown, she is at the top of the bump and is moving with a speed of 5 m/s.

Is the normal force exerted on the skateboarder by the bump *greater than*, *less than*, or *equal to* the weight of the skateboarder?

Explain.

NT5E-CT53: BLOCK HELD ON SMOOTH RAMP—WEIGHT AND NORMAL FORCE

A block is tethered to a frictionless ramp by a horizontal string as shown. The block is at rest.

Is the normal force exerted on the block by the ramp *greater than, less than,* **or** *equal to* **the weight of the block?**

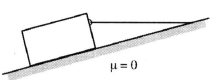

Explain.

NT5F-WWT54: TWO BLOCKS AT REST—NORMAL FORCE

In the situation on the left the block is sitting on a horizontal surface, and in the situation on the right, an identical block is sitting on a rough incline. A student comparing the normal force exerted on the block by the surface in the two cases states:

"*Since both blocks are identical, I think the normal forces are the same because in each case the normal force will be equal to the weight.*"

What, if anything, is wrong with this contention? If something is wrong, identify it, and explain how to correct it. If this contention is correct, explain why.

NT5F-RT55: Boxes on Rough Vertical Surface—Normal Force on Wall

Boxes are held at rest against rough vertical walls by forces pushing horizontally on the boxes as shown.

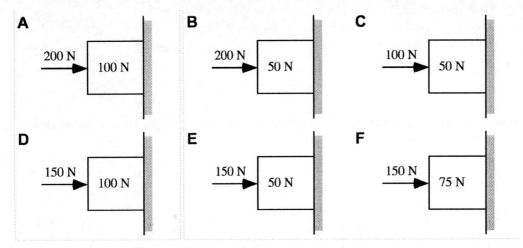

Rank the magnitude of the normal force exerted on the walls by these boxes.

    Greatest   1 _____ 2 _____ 3 _____ 4 _____ 5 _____ 6 _____ Least

OR, The magnitude of the normal force exerted on the wall by these boxes is the same
    but not zero.   ____

OR, The magnitude of the normal force exerted on the wall by these boxes is zero.   ____

OR, The ranking for the magnitudes of the normal forces cannot be determined.   ____

Explain your reasoning.

NT5F-CCT56: PULLING BOX OVER ROUGH HORIZONTAL SURFACE—NORMAL FORCE BY SURFACE

A rope that makes an angle of 30° with the horizontal is attached to a 50 N box that is moving along the floor. The force applied by the rope is 40 N. The coefficient of static friction between the box and the floor is 0.6 and the coefficient of kinetic friction is 0.4. Four students are discussing the normal force exerted on the box by the rough floor for this situation:

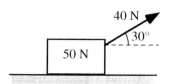

Aiko: *"The normal force is 50 N since that is the weight of the box. The normal force makes a Newton's Third Law pair with the weight."*

Bahir: *"But in this case there is also an upward force of 40 N by the rope. The normal force is only 10 N."*

Chloe: *"Actually it is only a part of that 40 N force that is acting upward. We'd have to use trigonometry to figure out how much, and then subtract this from the weight to get the normal force."*

Delbert: *"We can't figure out the normal force until we know the acceleration. The greater the acceleration, the less the normal force will be."*

Which, if any, of these students do you think is right?

Aiko _____ Bahir_____ Chloe _____ Delbert _____ **None of them** _____

Explain your reasoning.

NT5F-QRT57: STACKED BLOCKS—NORMAL FORCES

A student pushes two blocks, *A* and *B*, across a desk at a constant speed. The force exerted on block *A* by the student is directed horizontally to the left. The mass of block *A* is greater than the mass of block *B*.

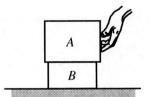

1. The magnitude of the normal force exerted on block *A* by block *B*

(a) is *greater than* magnitude of the normal force exerted on block *B* by block *A*.
(b) is *less than* magnitude of the normal force exerted on block *B* by block *A*.
(c) is *equal to* magnitude of the normal force exerted on block *B* by block *A*.
(d) *cannot be compared* to the magnitude of the normal force exerted on block *B* by block *A* based on the information given.

Explain.

2. The magnitude of the normal force exerted on block A by block B

(a) is *greater than* magnitude of the weight of block *A*.
(b) is *less than* magnitude of the weight of block *A*.
(c) is *equal to* magnitude of the weight of block *A*.
(d) *cannot be compared* to magnitude of the weight of block *A* based on the information given.

Explain.

3. The magnitude of the normal force exerted on block B by block A

(a) is *greater than* magnitude of the weight of block *B*.
(b) is *less than* magnitude of the weight of block *B*.
(c) is *equal to* magnitude of the weight of block *B*.
(d) *cannot be compared* to magnitude of the weight of block *B* based on the information given.

Explain.

4. The magnitude of the normal force exerted on block B by block A

(a) is *greater than* the magnitude of the normal force exerted on block *B* by the desk.
(b) is *less than* the magnitude of the normal force exerted on block *B* by the desk.
(c) is *equal to* the magnitude of the normal force exerted on block *B* by the desk.
(d) *cannot be compared* to the magnitude of the normal force exerted on block *B* by the desk based on the information given.

Explain.

NT5G-RT58: System of Accelerating Blocks—Tension

A system of two blocks connected by a massless string is pulled so that the system is accelerating. There is no friction between the blocks and the surfaces they move along. Values for six variations (labeled *A-F*) of this system that have different masses and accelerations are given in the table.

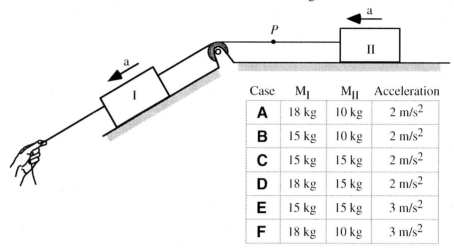

Case	M_I	M_{II}	Acceleration
A	18 kg	10 kg	2 m/s^2
B	15 kg	10 kg	2 m/s^2
C	15 kg	15 kg	2 m/s^2
D	18 kg	15 kg	2 m/s^2
E	15 kg	15 kg	3 m/s^2
F	18 kg	10 kg	3 m/s^2

Rank these situations on the basis of the tension in the string at point *P*.

Greatest 1 _____ 2 _____ 3 _____ 4 _____ 5 _____ 6 _____ Least

OR, The tension at point *P* is the same but not zero for all these systems. ____

OR, The tension at point *P* is zero for all these systems. ____

OR, We cannot determine the ranking for the tension at point *P* in these systems. ____

Please explain your reasoning.

NT5G-RT59: CART AND BLOCK—MASS OF HANGING BLOCK

The six figures below show systems of a block connected by a string to a cart so that they move together. The carts on the horizontal surface are identical in all six situations, but the hanging blocks are not. The instantaneous velocity and acceleration of each cart is given. The carts move without friction and the pulleys are massless and frictionless.

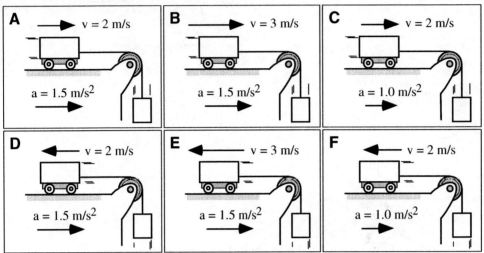

Rank these situations on the basis of the mass of the hanging block.

Greatest 1 _____ 2 _____ 3 _____ 4 _____ 5 _____ 6 _____ Least

OR, The hanging blocks all have the same mass.   ___

OR, We cannot determine the ranking for the masses of the hanging blocks.   ___

Please explain your reasoning.

nT5G-CT60: Hanging Mirror—Tension

Shown below is a mirror that has been hung by three strings from two nails on a wall.

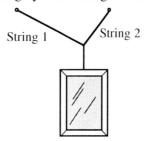

Will the tension in String 1 be *greater than*, *less than*, or *equal to* the tension in String 2?

Explain.

nT5G-WWT61: Blocks on a Rough Incline—Tension

Three identical blocks are tied together with ropes and pulled up a rough incline with an acceleration $\vec{a}$. A student who is asked to compare the tension in the rope at point P to the tension at point S states:

> "Each rope is pulling one block. All three blocks are accelerating at the same rate and they are identical. I think the tensions at points P and S will be the same."

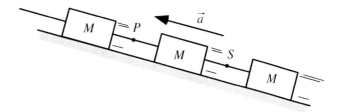

What, if anything, is wrong with this contention? If something is wrong, identify it, and explain how to correct it. If this contention is correct, explain why.

NT5G-CCT62: PULLING A CRATE ACROSS FLOOR—APPLIED FORCE

Grace pulls a large crate across a floor at a constant speed of 1.48 meters per second. Three supervisors are helping Grace by analyzing her efforts. Here is a snippet of their conversation:

Abelardo: *"All I know is Grace better not stop, or she'll never get the crate moving. By Newton's Third Law, for every action there is an equal and opposite reaction. So when Grace pulls on the crate, there will be an equal and opposite force on Grace. The net force will be zero, and so the crate will stay at rest."*

Bethanne: *"If Grace tipped the crate on its end, there would be less surface area in contact with the floor, so there would be less friction, and she wouldn't have to pull so hard."*

Craig: *"If Grace slowed down to 1 meter per second, she wouldn't have to pull so hard."*

Which, if any, of these supervisors do you agree with?

Abelardo _____ Bethanne _____ Craig _____ None of them _____

Please explain your reasoning.

NT5G-CCT63: THREE TEAM TUG-OF-WAR—FORCE COMPARISON

Three teams are engaged in a three-way tug-of-war. Each team is pulling on a rope that is attached to a tire. At the instant shown, the tire is moving to the west at a constant speed of 1 m/s. Three students are discussing the relative tensions in the ropes.

Amber: *Team 3 is winning. They must be pulling the hardest, so the tension in their rope is biggest.*

Bartolo: *Team 1 is fighting both teams. They've got to pull hard enough to the east to counteract Team 3, and hard enough to the north to counteract Team 2. So if Team 3 is pulling with a force of 150 pounds, and if Team 2 is pulling with a force of 100 pounds, then Team 1 has to pull with a force of 250 pounds. Maybe just a bit less, since they are losing.*

Chumlee: *There is no way we can tell which tension is biggest in this case. We'd need a lot more information than just the directions of the forces to know.*

Which, if any, of these students do you agree with?

Amber _____ Bartolo _____ Chumlee _____ None of them _____

Please explain your reasoning.

NT5G-WWT64: BLOCKS ON A ROUGH INCLINE—TENSION

Two identical blocks are tied together with a rope and pulled up a rough incline with an acceleration $\vec{a}$. A student considering the forces acting in this situation contends:

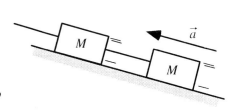

"I think the tension in the rope connecting the boxes has to be larger than the force the lower box exerts on that rope because the tension is causing the lower block to accelerate up the incline."

What, if anything, is wrong with this contention? If something is wrong, identify it, and explain how to correct it. If this contention is correct, explain why.

NT5G-QRT65: CHILD ON A SWING—TENSION

A child is swinging back and forth on a tire swing that is attached to a tree branch by a single rope. Shown are two positions during a swing from right to left. Three students are discussing the tension in the rope at the bottom of the swing.

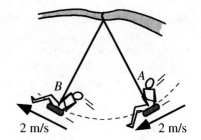

Alia: *"At the bottom of the swing, she will be moving exactly horizontally. Since she is not moving vertically at that instant, the vertical forces cancel. The tension in the rope at that instant equals the weight."*

Brian: *"Just looking at the velocity vectors, the change in velocity points upward between A and B. So that is the direction of the acceleration, and also of the net force. To get a net force pointing upward, the tension would have to be greater than the weight."*

Clara: *"But there aren't just two forces acting on her at the bottom of the swing. Since she's moving in a circle, there's also the centripetal force, which acts toward the center of the circle. Since both the tension and the centripetal force point upward, and the weight points downward, to get zero net force the tension actually has to be less than the weight. The tension plus the centripetal force equals the weight."*

Which, if any, of these students do you agree with?

Alia _____ Brian _____ Clara _____ None of them_____

Please explain your reasoning.

NT5G-CT66: PULLING A CRATE ACROSS FLOOR—APPLIED FORCE

In both cases below, Grace pulls a large crate across a floor at a constant speed of 1.48 meters per second.

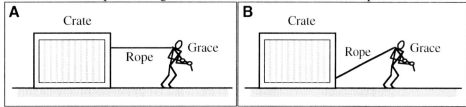

Is the magnitude of the force exerted by Grace on the rope in Case A *greater than, less than,* **or** *equal to* **the magnitude of the force exerted by Grace on the rope in Case B?**

Explain your reasoning.

NT5G-WWT67: HANGING STONE AND SLIDING BOX—FREE-BODY DIAGRAMS

A massless rope connects a box on a horizontal surface and a hanging stone as shown below. The rope passes over a massless, frictionless pulley. The box is given a quick tap so that it slides to the right along the horizontal surface. The figure below shows the block after it has been pushed while it is still moving to the right. The mass of the hanging stone is larger than the mass of the box. There is friction between the box and the horizontal surface. Free-body diagrams that a student has drawn to scale for the box and for the hanging stone are shown.

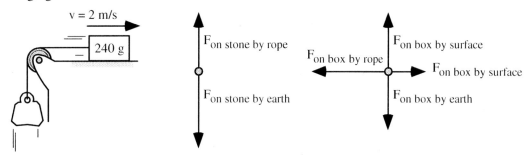

What, if anything, is wrong with these free-body diagrams? If something is wrong, identify it and explain how to correct it. If nothing is wrong, explain why the diagrams are appropriate.

NT5G-RT68: CART AND HANGING BLOCK—ACCELERATION

In each situation below, a block is attached by a string to a cart on a horizontal surface. The string passes over a massless, frictionless pulley. In C, D, and E the cart on the horizontal surface is tied to a fixed rod. In A, B, and F the carts move without friction. The masses of the blocks and of the carts are given in each figure.

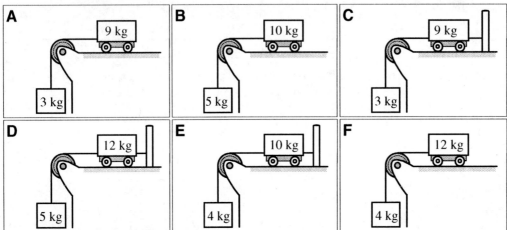

Rank these systems on the basis of the acceleration of the blocks.

       Greatest   1 _____ 2 _____ 3 _____ 4 _____ 5 _____ 6 _____ Least

OR, The acceleration is the same but not zero for all these systems.   ____

OR, The acceleration is zero for all these systems.   ____

OR, We cannot determine the ranking for the accelerations.   ____

Please explain your reasoning.

NT5G-RT69: WATER SKIERS—TENSION

In each of the six figures below, water skiers are being pulled at a constant speed by a towrope attached to a speedboat. Because the weight of the skiers and the type of skis they are using varies, they experience different resistive forces from the water. Values for this resistive force (RF) and for the speed of the skiers are given in each figure.

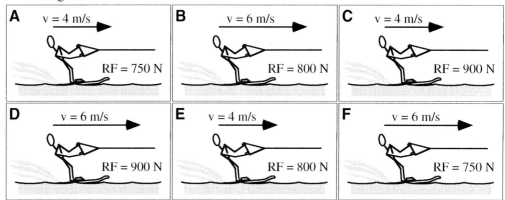

Rank these situations on the basis of the tension in the towrope.

Greatest   1 _____ 2 _____ 3 _____ 4 _____ 5 _____ 6 _____ Least

OR, The tension is the same but not zero for all these cases.   ____

OR, The tension is zero for all these cases.   ____

OR, We cannot determine the ranking for the tensions of these ropes.   ____

Please explain your reasoning.

NT5G-RT70: HANGING BLOCKS—TENSION

In each case shown, two blocks are supported by massless strings. The lower block is the same in all cases, but the mass of the upper block varies. The acceleration and velocity for each system is given in the figure.

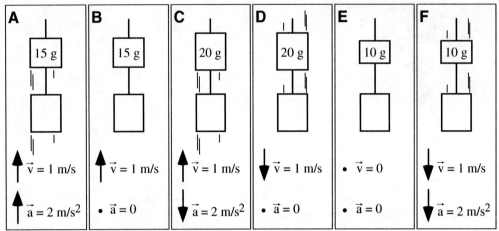

Rank these situations on the basis of the tension in the string between the blocks.

Greatest   1 _____ 2 _____ 3 _____ 4 _____ 5 _____ 6 _____ Least

OR, The tension in the string is the same for all these cases.   ____

OR, The tension in the string is zero for all these cases.   ____

OR, We cannot determine the rankings for the tensions here.   ____

Please explain your reasoning.

NT5G-CCT71: HANGING STONE CONNECTED TO BOX—FREE-BODY DIAGRAMS

A massless rope connects a box on a horizontal surface and a hanging stone as shown below. The rope passes over a massless, frictionless pulley. The box is given a quick tap so that it slides to the right along the horizontal surface. The figure below shows the block after it has been pushed while it is still moving to the right. The mass of the hanging stone is larger than the mass of the box. There is friction between the box and the horizontal surface. Free-body diagrams that a student has drawn to scale for the box and for the hanging stone are shown.

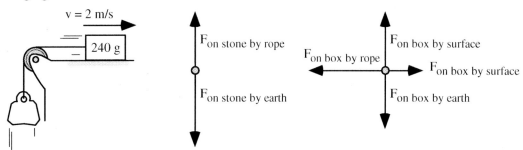

Four students discussing these free-body diagrams make the following contentions:

Ali: *"I think there is a problem with the free-body diagram for the hanging stone. The two forces should have the same magnitude."*

Brianna: *"But the stone is moving upward – there should be a larger force in that direction."*

Carlos: *"No, the diagram for the hanging stone is okay, but there is a problem with the diagram for the box. The frictional force is in the wrong direction."*

Dante: *"No, all three of you are wrong. Both free-body diagrams are correct because both show the way the objects would be accelerating."*

Which, if any, of these students do you think is right?

Ali _____ Brianna _____ Carlos _____ Dante _____ None of them _____

Explain your reasoning.

NT5G-CCT72: Carts Moving Along Horizontal Surface—Acceleration

In each case shown below, a cart is moving along a horizontal surface. The carts are the same size and shape but carry different loads, so their masses differ. All of the carts have a string attached, which passes over a pulley and is tied to a weight that is hanging free. All of the hanging weights are identical. As the carts move to the right, they will pull the hanging weights up toward the horizontal tabletop surface.

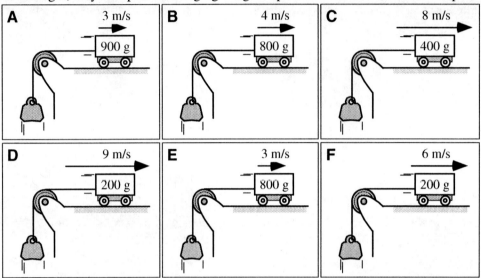

Four students are asked to rank these situations on the basis of the magnitude of the acceleration of the carts. They respond as follows:

Abran: *"I think the ranking should be D = F > C > B = E > A. Since the acceleration is inversely proportional to the mass and the net forces are the same, the differences in acceleration will depend on the mass only. The acceleration is causing each cart to slow down, so the lightest will be slowed down the quickest, while the heaviest will slow down the slowest."*

Bakari: *"No, the ranking is D > C > F > B > A > E, because the greater the initial speed, the greater the acceleration. If the speeds are the same then the mass will make the difference."*

Carol: *"I think the ranking should be D > F > C > B > E > A, since the carts traveling at a faster speed with a lesser mass will accelerate faster."*

Dolores: *"The acceleration of the carts are all zero because acceleration means an increase or decrease in speed. All of these objects are moving at a constant velocity and don't have any acceleration."*

Which, if any, of these students do you agree with?

Abran _____ Bakari _____ Carol _____ Dolores _____ None of them _____

Please explain your reasoning.

NT5G-RT73: HANGING STONE CONNECTED TO SLIDING BOX ON ROUGH SURFACE—ACCELERATION

In each case shown below, a box is sliding along a horizontal surface. There is friction between the box and the horizontal surface. The box is tied to a hanging stone by a massless rope running over a massless, frictionless pulley. All six cases are identical except for the different initial velocities of the boxes.

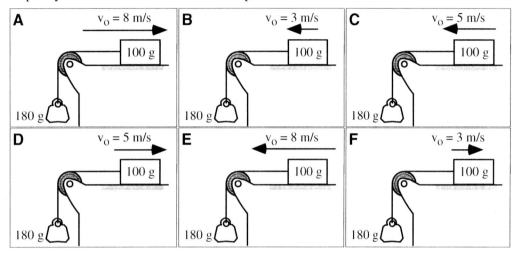

Rank the acceleration of these boxes at the instant shown.

Greatest   1 _____ 2 _____ 3 _____ 4 _____ 5 _____ 6 _____ Least

OR, The acceleration is the same but not zero of these boxes.   ____

OR, The acceleration is zero of these boxes.   ____

OR, We cannot determine the ranking for the acceleration of these boxes.   ____

Please explain your reasoning.

NT5G-CRT74: HANGING STONE CONNECTED TO A SLIDING BOX—VELOCITY VS. TIME

A box is sliding to the right along a horizontal surface with a velocity of 2 m/s. There is friction between the box and the horizontal surface. The box is tied to a hanging stone by a massless rope running over a massless, frictionless pulley. The mass of the stone is larger than the mass of the box. The box will slow down, come to rest at an instant, and then move to the left with increasing speed. Assume that a positive velocity represents motion to the right.

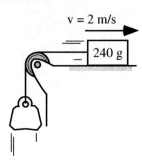

Which, if any, of the velocity versus time graphs below represent the movement of the sliding box?

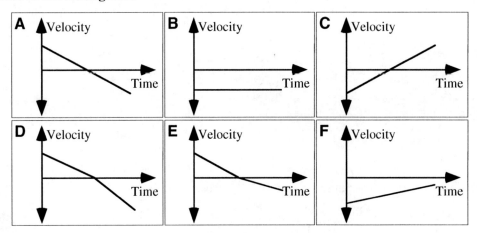

Explain.

NT5G-RT75: String Passing over a Pulley—Tension at Points

A student pulls on a massless string that passes over a frictionless pulley and is attached to a suspended mass. He is pulling the string horizontally so that, at the instant shown, the mass is moving upward at a constant speed.

Rank the tension at the labeled points *A*, *B*, *C*, and *D*.

Greatest   1 _____ 2 _____ 3 _____ 4 _____   Least

OR, The tension is the same but not zero for these points.   ___

OR, The tension is zero for these points.   ___

OR, We cannot determine the ranking for the tension at these points.   ___

Please explain your reasoning.

NT5G-CCT76: STRINGS CONNECTED TO A RING—TENSION IN STRINGS

The three strings in the arrangement shown are massless, and the mass of the ring can also be ignored. Strings 1 and 2 are the same length, and each makes an angle of 60° with the vertical. Four students make the following contentions:

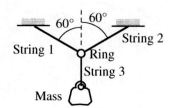

Arlene: *"The tension in strings 1 and 2 will each be half the weight of the suspended mass. Each holds up half of the weight."*

Bablu: *"The tension in strings 1 and 2 will each be greater than half the weight of the suspended mass. These strings have to fight each other as well as hold up the suspended mass."*

Cain: *"I think you have to calculate to decide. The tensions in strings 1 and 2 will depend on the sine of the angle, and since sine and cosine functions can never be greater than one the tension will be some fraction of the mass."*

Danyl: *"We can't compare the tension in strings 1 and 2 to the weight of the suspended mass unless we know the length of String 3. If string 3 is long enough, then to compensate the vectors for strings 1 and 2 will have to be really large."*

Which, if any, of these students do you agree with?

Arlene_____ Bablu _____ Cain _____ Danyl_____ None of them____

Explain.

NT5G-CCT77: String at Angle Passing over a Pulley—Tension at Points

A student holds a massless string that passes over a frictionless pulley and is attached to a suspended mass as shown. The mass is at rest. She then moves her hand so that the portion of string between her hand and the pulley moves from the horizontal to an angle as shown. Four students make the following contentions about this situation:

Aletheia: *"The tensions would all stay the same. The pulley will change the direction of the force, but the size of the force only depends on the mass, and that hasn't changed.*

Bem: *"Nothing has changed at point C, and the tension will stay the same there. But A and B are being pulled downward now as well as horizontally, so the tensions at A and B would increase."*

Charity: *"I agree with you that the tension at point C would stay the same, but I think the tensions at A and B would actually decrease. Now gravity is actually helping the hand to keep the mass in place, and so the hand won't have to pull as hard. Since A and B are close to where the hand is pulling, the tension there will go down.*

Dorothy: *I think the tension will increase at all three points. As the hand moves down so that it is more vertical, it's like it is fighting the pull of the mass more and more. So the string gets stretched tighter – the tension goes up all over.*

Which, if any, of these students do you agree with?

Aletheia_____ Bem _____ Charity _____ Dorothy _____ None of them_____

Explain.

NT5G-CCT78: UNEQUAL LENGTH STRINGS CONNECTED TO A RING—TENSION IN STRINGS

The three strings in the arrangement shown are massless, and the mass of the ring can also be ignored. String 1 is 1 meter long, and string 2 is 2 meters long. Four students make the following contentions:

Aba: *"The tension in strings 1 and 2 will each be half the weight of the suspended mass. They each have to hold up half of the weight."*

Belita: *"The tension in string 1 will be greater than the tension in string 2, because the string is shorter. The tension is more concentrated in a shorter string."*

Coco: *"String 1 is at twice the angle but has only half the length. These effects compensate, so the tension in string 1 is the same as the tension in string 2."*

David: *"The tension in string 2 will be greater than the tension in string 1 because this string is at a smaller angle to the vertical. When you add up the forces it will be the long side of the triangle."*

Which, if any, of these students do you agree with?

Aba_____ Belita _____ Coco _____ David _____ None of them_____

Explain.

NT5C-QRT79: MOTORCYCLIST SLOWING DOWN ON HILL—AVERAGE ACCELERATION & NET FORCE

At the instant shown, the motorcyclist is slowing down as she approaches a steeper portion of a hill.

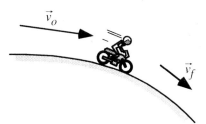

a) Use the velocity vectors shown to find the direction of the average acceleration of the rider and motorcycle (considered as a single system).

b) Draw a free-body diagram for the rider and motorcycle (considered as a single system), and show how the forces in your free-body diagram add to give a net force in the direction of the average acceleration.

NT5G-CT80: Ball Suspended from Ceiling by Two Strings—Tension

A 0.5-kg ball is suspended from a ceiling by two strings. The ball is at rest.

a) Is the tension in string 1 *greater than, less than,* or *the same as* the tension in string 2?

Explain.

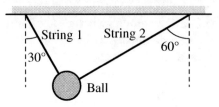

Suppose that the ceiling in the picture is the ceiling of an elevator, and that the elevator is moving *down* at a constant speed of 2 m/s.

b) Is the tension in string 1 *greater than, less than,* or *the same as* the tension in string 1 in the previous question (a) where the ball was at rest?

Explain.

NT5G-CT81: Gymnast Suspended by Two Ropes—Tension

A gymnast weighing 500 N is suspended by two ropes from the ceiling as shown. The gymnast is at rest.

Is the magnitude of the tension in the rope on the left *greater than, less than,* or *equal to* 250N?

Explain your reasoning.

NT5G-CT82: Mass Hanging Midway Between Two Walls—Tension

A hanging mass is suspended midway between two walls. The string attached to the left wall is horizontal while the string attached to the right wall makes an angle with the horizontal as shown. This angle (α) in Case A is larger than the angle (β) in Case B.

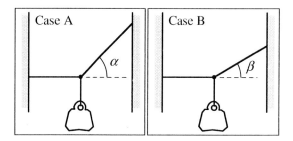

Is the tension in the horizontal string attached to the left wall *greater in Case A, greater in Case B,* or *the same in both cases*?

Explain your reasoning.

NT5G-CCT83: HANGING MASS—TENSION IN THREE STRINGS

A hanging mass is suspended midway between two walls. The string attached to the left wall is horizontal while the string attached to the right wall makes an angle with the horizontal as shown. This angle (α) in Case A is larger than the angle (β) in Case B. Four students make the following claims about the tensions in the strings:

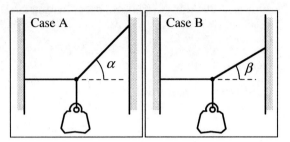

Abbie: *"I think the tensions in the three strings will all stay the same. The weight is the same, and the weight is still going to be divided up among the three ropes."*

Bobby: *"I think the tension in the horizontal and vertical strings are the same, because they are exactly the same in both cases. But in Case B the diagonal rope is shorter, so the tension is more concentrated there."*

Che: *"The diagonal string still has to hold the weight up by itself, because the horizontal string can't lift anything. So the diagonal string still has the same tension. But now it's pulling harder against the horizontal string because of the angle, so the tension in the horizontal string has to go up."*

Damian: *"But the diagonal string is fighting harder against the weight in Case A – it is pointing more nearly opposite the weight. So it has to have a greater tension in Case A. And since the tension in the diagonal string is greater, and the tension in the vertical string is the same, the tension in the horizontal string must be less in Case A. The tensions still have to balance out so that they are the same in both cases."*

Which, if any, of these students do you agree with?

Abbie _____ Bobby _____ Che _____ Damian _____ None of them _____

Explain.

NT5G-CCT84: TWO CONNECTED OBJECTS ACCELERATING DOWNWARD—TENSION

Two objects with masses of m_1 = 6 kg and m_2 = 10 kg are connected by a massless wire. They are pulled upward by an applied force F. The result is a constant acceleration of 3 m/s^2 downward for the two objects, because the force F is smaller than the total weight of the objects. The tension in the wire connecting the objects is T. Four students discuss this tension:

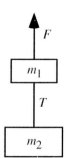

Anh: *"The tension is the net force on the lower object which is 30 N, since F_{net} = ma from Newton's Second Law. The lower object has a mass of 10 kg and it is accelerating at 3 m/s^2."*

Brandon: *"No, the tension in the wire is more than the net force of 30 N since the lower object has a weight of about 100 N. The tension should be 130 N since the 30 N, the net force, is added to 100 N, the weight."*

Cathy: *"You are both wrong, the tension in the wire is upward and should be less than the weight since the system is accelerating downward. It should be 70 N by applying Newton's Second Law and taking into account the directions of the forces."*

Deshi: *"I am confused about this. I do not think we can answer it until we know which direction the system is moving. Is it moving upward or downward? Won't that make a big difference on the tension? Perhaps it depends on the coordinate system we use."*

Which, if any, of these students do you think is right?

Anh _____ Brandon _____ Cathy _____ Deshi _____ None of them _____

Explain your reasoning.

NT5G-RT85: HANGING MASS—STRING TENSION

In each case below, a string is attached to one or more identical blocks. All of the masses are at rest. The strings are so light that they can be considered to be massless. In cases A–E, the strings pass over pulleys that are frictionless and massless.

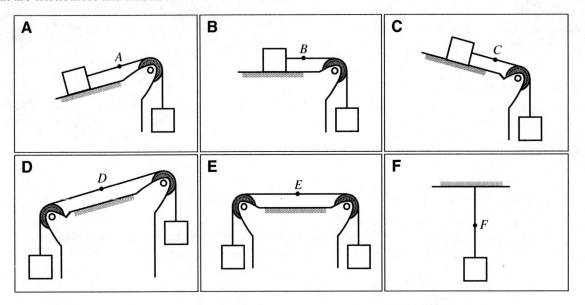

Rank these situations based on the tension in the strings at the labeled points.

Greatest   1 _____ 2 _____ 3 _____ 4 _____ 5 _____ 6 _____ Least

OR, The tension is the same for all these arrangements.   ____

OR, The tension is zero for all these arrangements.   ____

OR, We cannot determine the ranking for the tensions in these arrangements.   ____

Please explain your reasoning.

NT5G-WWT86: BLOCKS ON A SMOOTH INCLINE—TENSION

Three identical blocks are tied together with ropes and pulled up a smooth (frictionless) incline with an acceleration $\vec{a}$. A student who is asked to compare the tension in the rope at point P to the tension at point S states:

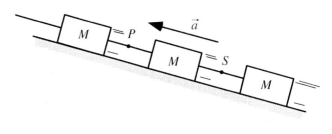

> *"Each rope is pulling one block. All three blocks are accelerating at the same rate and they are identical. I think the tensions at points P and S will be the same."*

What, if anything, is wrong with this contention? If something is wrong, identify it, and explain how to correct it. If this contention is correct, explain why.

NT5G-CCT87: BLOCKS ON A SMOOTH INCLINE—TENSION

Two identical blocks are tied together with ropes and pulled up a smooth (frictionless) incline with an acceleration $\vec{a}$. Three students are comparing the tension in the rope between the blocks to the force that the lower block exerts on that rope:

Albert: *"I think the tension has to be larger because it is causing the lower block to accelerate up the incline. If it was the same then the block wouldn't accelerate."*

Benifacio: *"I disagree. Force equals mass times acceleration, and the accelerations are the same. The rope hardly weighs anything compared to the block, so it can't exert as much force. The force the block exerts has to be greater."*

Connie: *"I agree that the rope and the block have exactly the same acceleration since they are moving together. But I think that means that the force has to be the same."*

Which, if any, of these three students do you agree with?

Albert_____ Benifacio _____ Connie _____ None of them_____

Explain.

NT5G-LMCT88: Two Connected Objects Accelerating Downward—Tension in Wire

Two objects with masses of $m_1 = 6$ kg and $m_2 = 10$ kg are connected by a massless wire. They are pulled upward by an applied force F. The result is a constant acceleration of 3 m/s² downward for the two objects, because the force F is smaller than the total weight of the objects. The tension in the wire between the objects is labeled T.

Identify from choices (a)-(e) how each change described below will affect the magnitude of the tension (T) in the wire between the objects.

Compared to the case above, this change will:

    (a) *increase* the magnitude of the tension in the wire.

    (b) *decrease* the magnitude of the tension in the wire but not to zero.

    (c) *decrease* the magnitude of the tension in the wire to zero.

    (d) *have no effect* on the magnitude of the tension in the wire.

    (e) *have an indeterminate* effect on the magnitude of the tension in the wire.

All of these modifications are the only changes to the initial situation shown in the diagram.

1) The mass of m_1 is decreased to 5 kg and the mass of m_2 is increased to 11 kg. _____

    Explain.

2) The mass of m_1 is increased to 7 kg and the mass of m_2 is decreased to 9 kg. _____

    Explain.

3) The applied force F is increased and the acceleration is 2 m/s² downward. _____

    Explain.

4) The applied force F is increased and the acceleration is 4 m/s² upward. _____

    Explain.

5) The applied force F is decreased and the acceleration is 4 m/s² downward. _____

    Explain.

6) The applied force F is decreased and the mass of m_1 is decreased. _____

    Explain.

NT5G-CT89: BLOCKS MOVING AT CONSTANT SPEED—TENSION IN CONNECTING STRING

Two identical blocks, *1* and *2*, are connected by a massless string as shown. In Case A, a student pulls on a string attached to block *2* so that the blocks travel to the right across a desk at a constant speed of 10 cm/s. In Case B, the student pulls on a string attached to block 1 so that the same blocks travel across the same desk to the left at a constant speed of 20 cm/s.

Will the tension in the diagonal string connecting the two blocks be *greater in Case A,* *greater in Case B,* **or** *the same in both cases?*

Please explain.

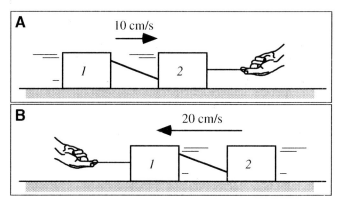

NT5G-WBT90: NEWTON'S SECOND LAW EQUATIONS—PHYSICAL SITUATION

Given below are two equations resulting from the application of Newton's Laws to a system of two objects:

$$Mg - T = Ma$$

$$T + \mu mg = ma$$

Describe and draw a physical situation that could have produced these equations.

NT5H-WBT91: NEWTON'S SECOND LAW EQUATION—PHYSICAL SITUATION

The equation below results from the application of Newton's Laws to an object:

$$27\ \text{N} - (\mu)(14\ \text{kg})(9.8\ \text{m/s}^2) = 0$$

Draw a physical situation that would result in this equation, and explain how your drawing is consistent with the equation.

NT5H-WBT92: NEWTON'S SECOND LAW EQUATION—PHYSICAL SITUATION

The equation below results from the application of Newton's Laws to an object:

$$57\ \text{N} - (\mu)(14\ \text{kg})(9.8\ \text{m/s}^2)(\cos 18°) - (14\ \text{kg})(9.8\ \text{m/s}^2)(\sin 18°) = 0$$

Draw a physical situation that would result in this equation, and explain how your drawing is consistent with the equation.

NT5H-RT93: BLOCK AT REST—STATIC FRICTIONAL FORCE

The figures below show six situations where the same block, which has a mass of 5 kg, remains at rest on either a horizontal or an inclined surface. The surfaces are all made of the same material. In all cases except Case A, a 4 N force acts on the block parallel to the surface as indicated by the arrow in the diagram.

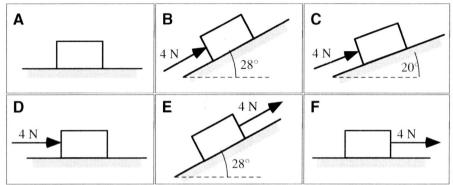

Rank these situations on the basis of the magnitude of the frictional force on the block.

Greatest   1 _____ 2 _____ 3 _____ 4 _____ 5 _____ 6 _____ Least

OR, The magnitude of the frictional force is the same but not zero for all these cases.   ____

OR, The magnitude of the frictional force is zero for all these cases.   ____

OR, We cannot determine the ranking for the magnitude of these frictional forces.   ____

Please explain your reasoning.

NT5H-CCT94: Box Pulled on Rough Horizontal Surface—Frictional Force on Box

A 100 N box is initially at rest on a rough horizontal surface. The coefficient of static friction is 0.6 and the coefficient of kinetic friction is 0.4. A constant 35 N horizontal force to the right is applied to the box as shown. Several students are discussing the frictional force exerted on the box by the rough surface 1 second after the force is first applied:

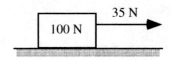

Al: *"The frictional force is 60 N since the box will not be moving and the coefficient of static friction is 0.6 with a normal force of 100 N."*

Brianna: *"The frictional force is 40 N since the coefficient of kinetic friction is 0.4 and there is a normal force of 100 N."*

Carlos: *"The frictional force is 35 N since the box will not be moving and the frictional force will cancel out the applied force of 35 N."*

David: *"It is 40 N for the kinetic frictional force and 60 N for the static frictional force. The normal force is 100 N and the coefficient of kinetic friction is 0.4 giving 40 N for the kinetic friction. Similarly, for the static frictional force it is 60 N since it has a coefficient of static friction of 0.6."*

Which, if any, of these students do you think is right?

Al _____ Brianna _____ Carlos _____ David _____ None of them _____

Explain your reasoning.

NT5H-CT95: Force on Box Moving over Horizontal Surface—Frictional Force on Box

In both cases below, a moving 50 N box has a force on it of 40 N that makes an angle of 30° with the horizontal. The coefficient of static friction between the box and the rough surface is 0.6 and the coefficient of kinetic friction is 0.4.

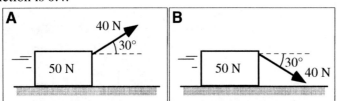

Will the frictional force exerted on the box by the rough surface in Case A be *greater than*, *less than*, or *equal to* the frictional force on the box by the rough surface in Case B?

Explain.

NT5H-LMCT96: Box Pulled on Rough Horizontal Surface—Frictional Force on Box

A 100 N box is initially at rest on a rough horizontal surface. The coefficient of static friction between the box and the surface is 0.6 and the coefficient of kinetic friction is 0.4. A constant 35 N force is applied to the box horizontally as shown.

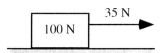

Identify from choices (a)-(e) how each change described below will affect the frictional force on the box by the surface *1 second after the horizontal force is first applied*.

Compared to the case above, this change will:

    (a) *increase* the frictional force exerted on the box by the surface.

    (b) *decrease* the frictional force exerted on the box by the surface but not to zero.

    (c) *decrease* the frictional force exerted on the box by the surface to zero.

    (d) *have no effect* on the frictional force exerted on the box by the surface.

    (e) *have an indeterminate* effect on the frictional force exerted on the box by the surface.

All of these modifications are changes to the initial situation shown in the diagram.

1) The weight of the box is changed to 50 N. _____

2) The weight of the box is changed to 200 N. _____

3) The applied force is increased to 50 N. _____

4) The applied force is increased to 80 N. _____

5) The coefficient of static friction is increased to 0.7. _____

6) The coefficient of kinetic friction is increased to 0.5. _____

7) The coefficient of kinetic friction is increased to 0.5 and the coefficient of static friction is increased to 0.7. _____

8) The weight of the box is changed to 200 N and the coefficient of static friction is increased to 0.7. _____

9) The weight of the box is changed to 200 N and the coefficient of kinetic friction is increased to 0.5. _____

10) The weight of the box is changed to 200 N and the applied force is increased to 50 N. _____

A 100 N box is initially at rest on a rough horizontal surface. The coefficient of static friction between the block and the surface is 0.6 and the coefficient of kinetic friction is 0.4. A man decides to move the box so he applies a steadily increasing horizontal force $F_{Applied}$ on the box to the right as shown. The graph on the left below shows the horizontal force $F_{Applied}$ on the box as it goes from 0 to 80 N as a function of time over 4 seconds.

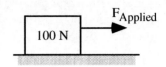

In the graph on the right below, sketch the frictional force exerted on the box by the surface as a function of time.

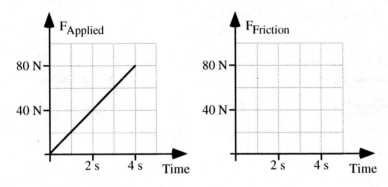

Explain.

NT5H-CCT98: Box on Rough Vertical Surface—Frictional Force on Box

A woman is applying a horizontal force on a 100 N box to the right (positive *x*-direction) to hold it in place against a rough vertical surface. The coefficient of static friction between the box and the surface is 0.6 and the coefficient of kinetic friction is 0.4. Several students are discussing the frictional force on the box 1 second after she first applies a constant horizontal force of 200 N:

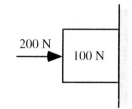

Ari: *"The frictional force is 60 N since the box will not be moving and the coefficient of static friction is 0.6."*

Bratislav: *"The frictional force is 100 N upward since the box has a weight of 100 N downward."*

Celeste: *"The frictional force will be 120 N since the box will not be moving and the normal force will be 200 N."*

Deshi: *"The frictional force will be 40 N for the kinetic frictional force and 60 N for the static frictional force. The weight is 100 N and the coefficient of kinetic friction is 0.4 giving 40 N for the kinetic friction. Likewise, for the static frictional force it has a coefficient of static friction of 0.6 giving a static frictional force of 60 N."*

Which, if any, of these students do you think is right?

Ari _____ Bratislav _____ Celeste _____ Deshi _____ None of them _____

Explain your reasoning.

NT5H-RT99: BOXES ON ROUGH VERTICAL SURFACE—FRICTIONAL FORCES ON THE WALL

In each case below, a box is held at rest against a rough vertical surface by a force pushing horizontally as shown. Values for the applied force and the weight of the boxes are given. The boxes are all made of the same material and the walls are identical.

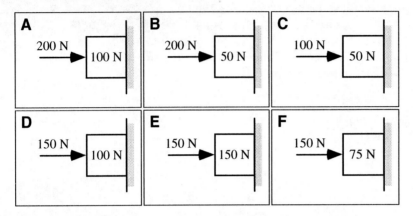

Rank the magnitude of the frictional force exerted on the wall by these boxes.

Greatest   1 _____ 2 _____ 3 _____ 4 _____ 5 _____ 6 _____ Least

OR, The magnitude of the frictional force exerted on the wall by all these boxes is the same but not zero. ____

OR, The magnitude of the frictional force exerted on the wall by all these boxes is zero. ____

OR, The ranking for the magnitude of the frictional force on the wall by these boxes cannot be determined. ____

Explain your reasoning.

NT5H-QRT100: Moving Stacked Blocks—Friction Forces

A student pushes two blocks, *A* and *B*, across a desk at a constant speed of 20 cm/s. The force exerted on block *A* by the student is directed horizontally to the left. The mass of block *A* is greater than the mass of block *B*.

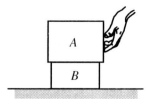

1. The magnitude of the friction force exerted on block *A* by block *B*

(a) is *greater than* the magnitude of the friction force exerted on block *B* by block *A*.
(b) is *less than* the magnitude of the friction force exerted on block *B* by block *A*.
(c) is *equal to* the magnitude of the friction force exerted on block *B* by block *A*.
(d) cannot be compared to the magnitude of the friction force exerted on block *B* by block *A* based on the information given.

Explain.

2. The magnitude of the friction force exerted on block *B* by the desk

(a) is *greater than* the magnitude of the friction force exerted on block *B* by block *A*.
(b) is *less than* the magnitude of the friction force exerted on block *B* by block *A*.
(c) is *equal to* the magnitude of the friction force exerted on block *B* by block *A*.
(d) cannot be compared to the magnitude of the friction force exerted on block *B* by block *A* based on the information given.

Explain.

3. The magnitude of the friction force exerted on block *A* by block *B*

(a) is *greater than* the magnitude of the force exerted on block *A* by the hand.
(b) is *less than* the magnitude of the force exerted on block *A* by the hand.
(c) is *equal to* the magnitude of the force exerted on block *A* by the hand.
(d) cannot be compared to the magnitude of the force exerted on block *A* by the hand based on the information given.

Explain.

NT5H-QRT101: Slowing Down Stacked Blocks—Friction Forces

A student pushes two blocks, *A* and *B*, across a desk. At the instant shown, the blocks are *slowing down*. The force exerted on block *A* by the student is directed horizontally to the left. The mass of block *A* is greater than the mass of block *B*.

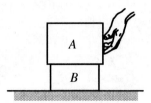

1. The magnitude of the friction force exerted on block *A* by block *B*

(a) is *greater than* the magnitude of the friction force exerted on block *B* by block *A*.
(b) is *less than* the magnitude of the friction force exerted on block *B* by block *A*.
(c) is *equal to* the magnitude of the friction force exerted on block *B* by block *A*.
(d) cannot be compared to the magnitude of the friction force exerted on block *B* by block *A* based on the information given.

Explain.

2. The magnitude of the friction force exerted on block *B* by the desk

(a) is *greater than* the magnitude of the friction force exerted on block *B* by block *A*.
(b) is *less than* the magnitude of the friction force exerted on block *B* by block *A*.
(c) is *equal to* the magnitude of the friction force exerted on block *B* by block *A*.
(d) cannot be compared to the magnitude of the friction force exerted on block *B* by block *A* based on the information given.

Explain.

3. The magnitude of the friction force exerted on block *A* by block *B*

(a) is *greater than* the magnitude of the force exerted on block *A* by the hand.
(b) is *less than* the magnitude of the force exerted on block *A* by the hand.
(c) is *equal to* the magnitude of the force exerted on block *A* by the hand.
(d) cannot be compared to the magnitude of the force exerted on block *A* by the hand based on the information given.

Explain.

NT5I-WWT102: Two Planetary Objects—Gravitational Force on Each

Two planetary objects with masses of m and $3m$ exert gravitational forces on each other. A student contends that the forces will be in different directions and of different magnitudes as shown.

$F_{\text{on A by B}}$　m　　　　　　　　　　$3m$　　$F_{\text{on B by A}}$

Planet A　　　　　　　　　　　　Planet B

What's wrong, if anything, with this student's contention? If something is wrong, identify it and explain how to correct it. If this student's contention is correct, explain why.

NT5I-QRT103: Three Objects Exerting Gravitational Forces—Net Force

Three objects each with a mass of M exert gravitational forces on each other. **Which of the arrows below show the direction of the net force on mass B?**

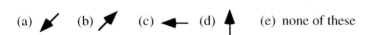

(a) ↙　(b) ↗　(c) ←　(d) ↑　(e) none of these

Explain.

NT5I-QRT104: Two Objects—Gravitational Force on Each

The drawing to the right shows an object (labeled B) that has mass m. To the left is another object (labeled A) that has mass $2m$. **Identify the pair of force vectors (the arrows) that correctly compare the gravitational force exerted on A by B with the gravitational force exerted on B by A.**

$2m$ m

○   ○

A   B

	Gravitational force on A by B	Gravitational force on B by A
(a)	⟶ (long)	⟵ (long)
(b)	⟶ (short)	⟵ (short)
(c)	⟶ (short)	⟵ (long)
(d)	⟵ (short)	⟶ (short)
(e)	⟵ (long)	⟶ (short)
(f)	⟵ (short)	⟶ (long)

Explain.

NT5J-QRT105: Rotating Cylinders—Force, Acceleration, & Velocity Directions

Four identical small cylinders rest on a circular horizontal turntable at the positions shown in the top view diagram at right. The turntable is rotating clockwise at a constant rate.

At the positions shown in the diagram, indicate the direction of the velocity, the acceleration, and the net force for each cylinder. Use the directions labeled on the rosette. Use the letter 'L' to indicate a direction into the page and the letter 'M' to indicate a direction out of the page.

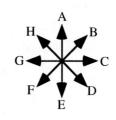

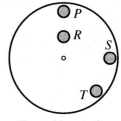

Turntable rotating
clockwise – top view

	Cylinder at P	Cylinder at R	Cylinder at S	Cylinder at T
(a) Direction of the velocity				
(b) Direction of the net force				
(c) Direction of the acceleration				

NT5J-RT106: Cars Going in Circles—Net Force

In each case below, a stock car travels around a circular track at a constant speed. The radii of the tracks and the speeds of the cars are given for each case.

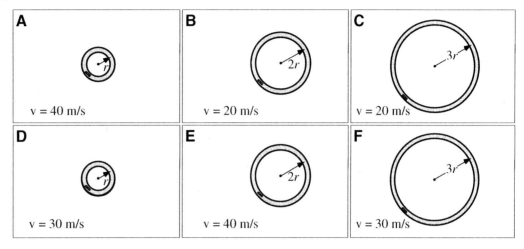

Rank the magnitude of the net force acting on the cars on these tracks.

Greatest   1 _____ 2 _____ 3 _____ 4 _____ 5 _____ 6 _____ Least

OR, The magnitude of the net force on the stock cars is the same but not zero for all these tracks. ____

OR, The magnitude of the net force on the stock cars is zero for all these tracks. ____

OR, We cannot determine the ranking for the magnitude of the net force on the stock cars. ____

Please explain your reasoning.

NT5J-CCT107: Ball Whirled in Vertical Circle—Net Force on Ball

A ball with a weight of 2 N is attached to the end of a cord of length 2 meters. The ball is whirled in a vertical circle counterclockwise as shown below. The tension in the cord at the top of the circle is 7 N and at the bottom it is 15 N. (Do not assume that the speed of the ball is the same at these points.)

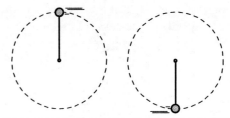

a) Three students discuss the net force on the ball at the top.

Angelica: *"The net force on the ball at the top position is 7 N since the net force is the same as the tension."*

Bo: *"The net force on the ball at the top position is 9 N. Both the tension and the weight are acting downward so you have to add them."*

Charles: *"No, you are both wrong. You need to figure out the centripetal force (mv^2/r) and include it in the net force."*

Which, if any, of these students do you think is right?

Angelica _____ Bo _____ Charles _____ None of them _____

Explain your reasoning.

b) Now the students discuss the net force on the ball at the bottom.

Angelica: *"The net force on the ball at the bottom position is 15 N since the net force is the same as the tension."*

Bo: *"The net force on the ball at the bottom position is 17 N, since you need to add the weight of 2 N to the tension of 15 N."*

Charles: *"The net force on the ball at the bottom position is 13 N. I agree that you need to take into account both the weight and the tension but they are in different directions so they will subtract."*

Which, if any, of these students do you think is right?

Angelica _____ Bo _____ Charles _____ None of them _____

Explain your reasoning.

NT5J-RT108: ROTATING CYLINDERS—NET FORCE MAGNITUDE

Four identical small cylinders rest on a circular horizontal turntable at the positions shown in the top view diagram at right. The turntable is rotating clockwise at a constant angular velocity.

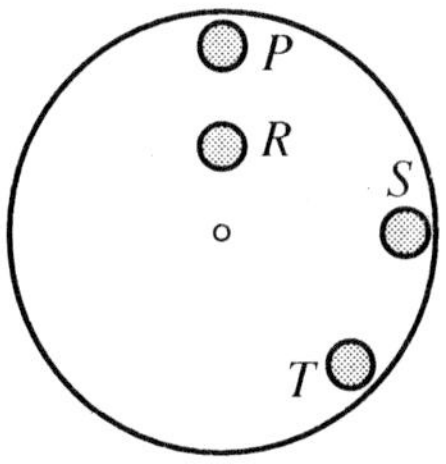

Rank the magnitude of the net force on the cylinder at the positions on the turntable indicated in the diagram.

Greatest   1 _____ 2 _____ 3 _____ 4 _____   Least

OR, The magnitude of the net force on the cylinder at these positions is the same but not zero. ____

OR, The magnitude of the net force on the cylinder at all of these positions is zero. ____

OR, The ranking for the magnitude of net force on the cylinder at these positions cannot be determined. ____

Explain your reasoning.

Turntable rotating clockwise – top view

NT6 WORK AND ENERGY

NT6A-WWT1: OBJECT CHANGING VELOCITY—WORK

A 2-kg object accelerates as a net force acts on it. During the 5 seconds this force acts, the object changes its velocity from 3 m/s east to 7 m/s west.

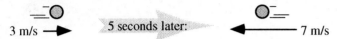

3 m/s ➡   5 seconds later: ⬅——— 7 m/s

A student states:

> "The change in kinetic energy of this object during these 5 seconds was 40 J, and thus the work done on this object by the net force during this period was also 40 J."

What, if anything, is wrong with this statement? If something is wrong, identify it, and explain how to correct it. If this statement is correct, explain why.

NT6A-CCT2: BICYCLIST ON A STRAIGHT ROAD—WORK

A bicyclist initially travels at a steady 8 m/s for 100 seconds on a straight level road, and then takes 40 seconds to slow to 5 m/s. Three students discussing this situation make the following contentions about the bicycle's kinetic energy:

Axel: *"The bicycle is just going to slow down naturally. It doesn't take any work for something to slow down."*

Bram: *"I disagree. The speed of the bike decreased, so there is a change in kinetic energy. That means work was done on the bike."*

Cassie: *"I think Axel is right that no work was done, but I don't agree with his reason. There is no work being done here because there are no external forces being exerted."*

Which, if any, of these students do you agree with?

Axel_____ Bram _____ Cassie _____ None of them_____

Explain.

NT6A-WWT3: Boat Position vs. Time Graphs—Work

Shown are graphs of the position versus time for two boats traveling along a narrow channel. The scales on both axes are the same for the graphs. In each graph, two points are marked with dots.

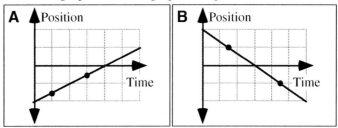

A student who is using these graphs to compare the net work done on the two boats between the two points says:

"I think that more net work was done on the boat in graph B because it moved farther during the interval between the points."

What, if anything, is wrong with this statement? If something is wrong, identify it, and explain how to correct it. If this statement is correct, explain why.

NT6A-BCT4: TUGBOAT CHANGING VELOCITY I—WORK & KINETIC ENERGY BAR CHART

a) The velocity of a tugboat increases from 2 m/s to 4 m/s in the same direction as a force is applied to the tugboat for 20 seconds.

Fill in the missing bars for the work & kinetic energy bar chart for this process.

2 m/s　　　　　　　4 m/s

Explain.

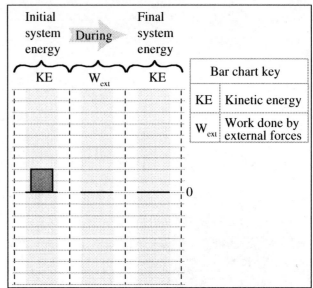

b) The velocity of a tugboat changes from 2 m/s to 4 m/s in the other direction as a force is applied to the tugboat for 20 seconds.

Fill in the missing bars for the work & kinetic energy bar chart for this process.

2 m/s　　　　　　　4 m/s

Explain.

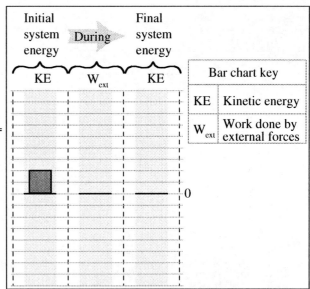

NT6A-BCT5: Tugboat Changing Velocity II—Work & Kinetic Energy Bar Chart

a) The velocity of a tugboat changes from 2 m/s west to 4 m/s west as a force is applied to the tugboat for 20 seconds.

Draw a work & kinetic energy bar chart for this process.

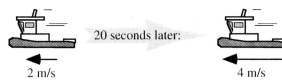

Explain.

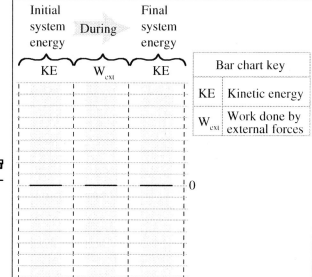

b) The velocity of a tugboat changes from 4 m/s west to 2 m/s west as a force is applied to the tugboat for 20 seconds.

Draw a work & kinetic energy bar chart for this process.

Explain.

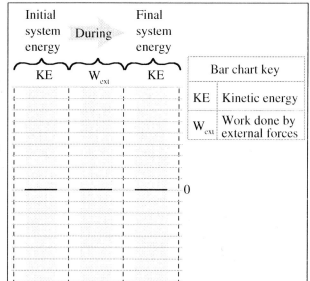

NT6A-BCT6: Box Pulled on Rough Surface—Work & Kinetic Energy Bar Charts

A 100 N box is initially at rest on a rough horizontal surface where the coefficient of static friction is 0.6 and the coefficient of kinetic friction is 0.4. A student decides to move the box by applying a horizontal force of 80 N to the box to the right as shown. The box starts at rest at point A.

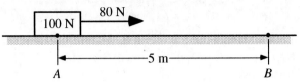

Fill in the work–kinetic energy bar charts below for the box as it moves between points A and B. The chart at left below has columns for work done by four kinds of external force. In the summary chart at right below you should include the *net* work done on the box as it moves from A to B.

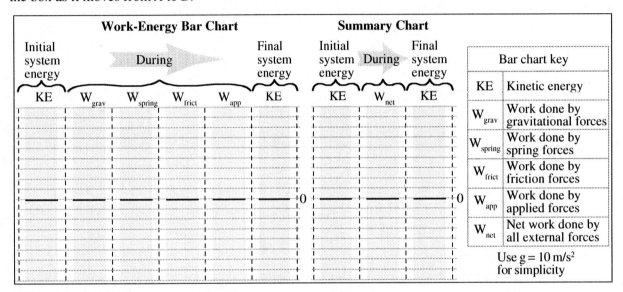

Explain.

NT6A-BCT7: Box Moving Upward I—Work & Kinetic Energy Bar Charts

A 100-N box is initially moving upward at 10 m/s. A student is applying a vertical force of 80 N upward with his hand as shown.

Fill in the work–kinetic energy bar charts below for the box as it moves upward a distance of 5 meters. The chart at left below has columns for work done by four kinds of external force. In the summary chart at right below you should include the *net* work done on the box as it moves upward 5 meters.

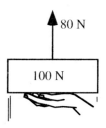

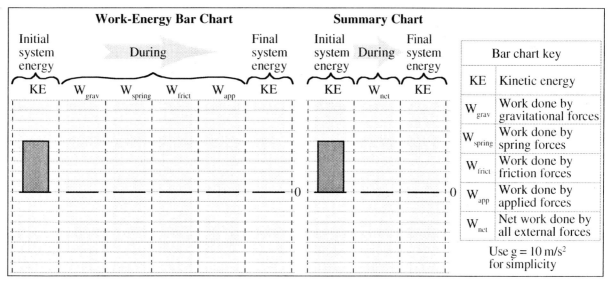

Explain.

NT6A-BCT8: Box Moving Upward II—Work & Kinetic Energy Bar Charts

A 100-N box is initially moving upward at 10 m/s. A student is applying a vertical force of 120 N upward with her hand as shown.

Fill in the work–kinetic energy bar charts below for the box as it moves upward a distance of 5 meters. The chart at left below has columns for work done by four kinds of external force. In the summary chart at right below you should include the *net* work done on the box as it moves upward 5 meters.

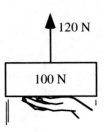

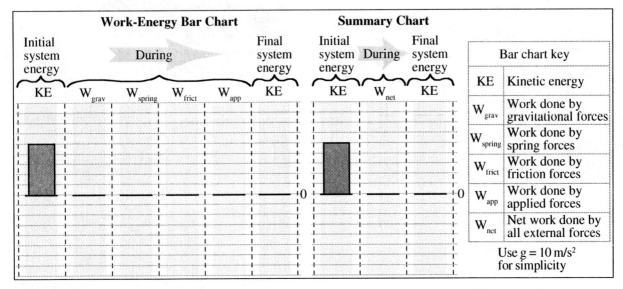

Explain.

NT6A-BCT9: Box Moving Downward—Work & Kinetic Energy Bar Charts

An 80-N box is initially moving downward at 10 m/s. A student is applying a vertical force of 120 N upward on the box with her hand as shown.

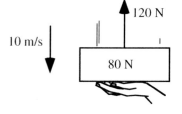

Fill in the work–kinetic energy bar charts below for the box as it moves downward a distance of 5 meters. The chart at left below has columns for work done by four kinds of external force. In the summary chart at right below you should include the *net* work done on the box as it moves downward 5 meters.

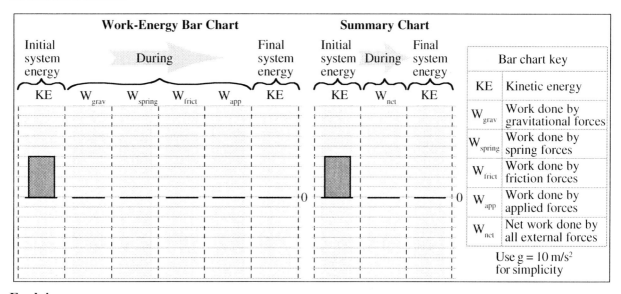

Explain.

NT6A-BCT10: BOX ATTACHED TO SPRING—WORK & KINETIC ENERGY BAR CHARTS

A 40-N box is initially at rest on a rough horizontal surface. A spring with spring constant 10 N/m connects the box to the wall and is unstretched. A 60 N force is applied horizontally to the right as shown. The maximum static friction force between the box and the surface is 15 N, and there is a kinetic friction force of 10 N between the box and the surface when the box is moving.

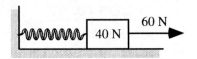

Fill in the work–kinetic energy bar charts below for the box as it moves to the right a distance of 4 meters. The chart at left below has columns for work done by four kinds of external force. In the summary chart at right below you should include the *net* work done on the box as it moves to the right 4 meters.

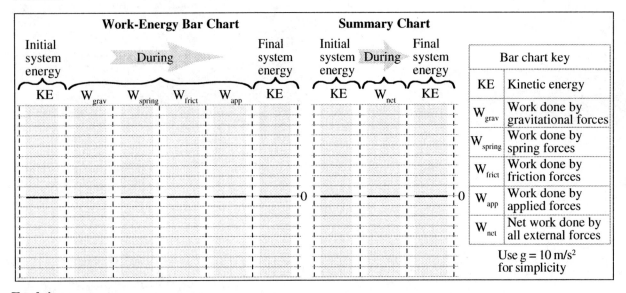

Explain.

NT6B-RT11: Toboggans on a Horizontal Surface—Speed

The figures below show identical toboggans that have traveled down a snowy hill. The toboggans all have the same speed at the bottom of the hill. Assume that the horizontal surfaces that they travel along are frictionless except for the shaded areas, where the coefficient of friction is given. These shaded areas have different lengths as shown.

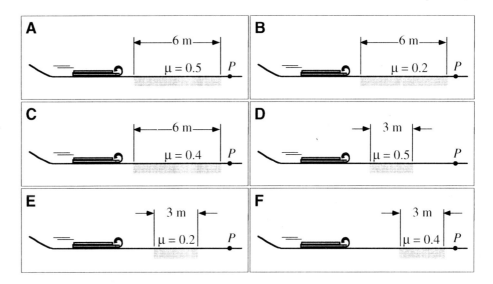

Rank these situations on the basis of the speed of the toboggans as they reach point *P*.

Greatest   1 _____ 2 _____ 3 _____ 4 _____ 5 _____ 6 _____ Least

OR, The speed is the same but not zero for all these toboggans.   ____

OR, The speed is zero for all these toboggans.   ____

OR, We cannot determine the ranking for the speed of these toboggans.   ____

Please explain your reasoning.

NT6B-LMCT12: BLOCK PUSHED ON INCLINE—WORK DONE

A block is pushed so that it moves up a ramp at constant speed.

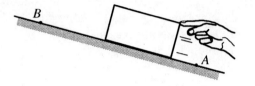

Identify from choices (a)-(e) below the appropriate description for the work done by the specified force while the block moves from point *A* to point *B*.

    (a) is *zero.*

    (b) is *less than* zero.

    (c) is *greater than* zero.

    (d) could be *positive or negative* depending on the choice of coordinate systems.

    (e) *cannot be determined.*

(1) The work done on the block by the hand   _____

Explain.

(2) The work done on the block by the normal force from the ramp   _____

Explain.

(3) The work done on the block by friction   _____

Explain.

(4) The work done on the block by the gravitational force   _____

Explain.

(5) The net work done on the block   _____

Explain.

NT6B-RT13: STACKED BLOCKS SETS—WORK TO ASSEMBLE

Shown below are five stacks, each containing three blocks. The masses of the blocks are given in the diagram in terms of M, the mass of the smallest block. Each block has the same height and has its center of mass at the center of the block. Originally, all the blocks were flat on the ground.

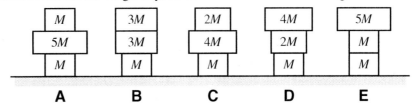

Rank the work required to assemble each stack.

Greatest   1 _____ 2 _____ 3 _____ 4 _____ 5 _____ Least

OR, The work required to assemble each stack is the same but not zero.   ____

OR, The work required to assemble each stack is zero.   ____

OR, The ranking for the work required to assemble the stacks cannot be determined.   ____

Explain your reasoning.

NT6B-RT14: Velocity vs. Time Graphs for Identical Objects—Work Done

Shown below are graphs of velocity versus time for six identical objects that move along a straight, horizontal, frictionless surface. A single external force acts on each object.

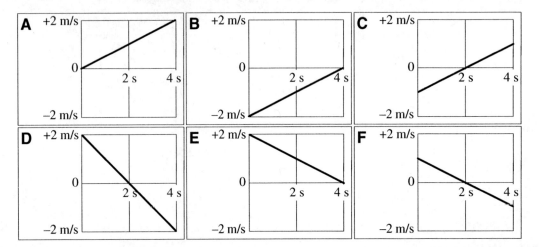

Rank the work done on the objects by the external force during the 4-second time interval shown.

Greatest 1 _____ 2 _____ 3 _____ 4 _____ 5 _____ 6 _____ Least

OR, The work done is the same but not zero for all these situations. ____

OR, The work done is zero for all these situations. ____

OR, We cannot determine the ranking for the work done for these situations. ____

Please explain your reasoning.

NT6B-RT15: Velocity vs. Time Graphs for Different Objects—Work Done

Shown below are graphs of velocity versus time for six objects that move along a straight, horizontal, frictionless path. A single external force acts on each object. The mass of each object is given.

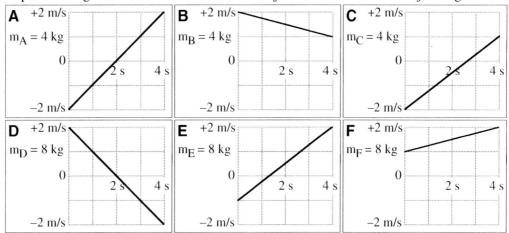

Rank the work done on the objects by the external force during the 4-second time interval shown.

Greatest   1 _____ 2 _____ 3 _____ 4 _____ 5 _____ 6 _____ Least

OR, The work done is the same but not zero for all these situations.   ____

OR, The work done is zero for all these situations.   ____

OR, We cannot determine the ranking for the work done for these situations.   ____

Please explain your reasoning.

NT6B-CCT16: Blocks Sliding Down Frictionless Ramps—Work by the Normal Force

Two identical blocks are released from rest at the same height. Block A slides down a steeper ramp than Block B. Both ramps are frictionless. The blocks reach the same final height indicated by the lower dashed line. Three students comparing the work done on the two blocks by the normal force state:

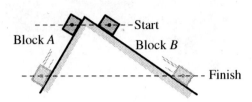

Annika: *"I think the normal force doesn't do any work on either block. The force on the block by the ramp is perpendicular to the ramp, and the displacement is parallel to the ramp. So the dot product is zero."*

BoBae: *"Work is force times displacement. The work done on Block A is negative, while the work done on Block B is positive, because the displacement for B is in the positive direction, while the displacement for A is in the negative direction."*

Craig: *"Since work is force times distance, and the distance the block travels is greater for Block B, the work done is greater for Block B."*

Which, if any, of these students do you agree with?

Annika _____ BoBae _____ Craig _____ None of them_____

Please explain your reasoning.

NT6B-CCT17: BLOCKS SLIDING DOWN FRICTIONLESS RAMPS—WORK BY THE EARTH

Two identical blocks are released from rest at the same height. Block *A* slides down a steeper ramp than Block *B*. Both ramps are frictionless. The blocks reach the same final height indicated by the lower dashed line. Three students are comparing the work done on the two blocks by the gravitational force (the weight of the blocks):

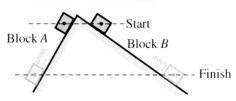

Asmita: *"Work is the dot product of force and displacement, and the weight is the same since the blocks are identical. But Block B travels further, so more work is done on Block B by the gravitational force than on Block A."*

Ben: *"Both blocks fall the same vertical distance, so the work done is the same."*

Cocheta: *"By Newton's Third Law, the force exerted on the block by the earth is exactly cancelled by the force exerted on the earth by the block. The work done is zero."*

Danae: *"The dot product depends on the angle that the force makes with the displacement. If we put the displacement and force vectors tail-to-tail, the angle is smaller for Block B than for Block A, and so the work done is greater."*

Which, if any, of these students do you agree with?

Asmita _____ Ben _____ Cocheta _____ Danae _____ None of them_____

Please explain your reasoning.

NT6B-QRT18: Blocks Sliding Down Frictionless Inclines—Work and Energy

Two identical blocks are released from rest at the same height at the same time. Block A slides down a steeper ramp than Block B. Both ramps are frictionless. The blocks reach the same final height indicated by the lower dashed line.

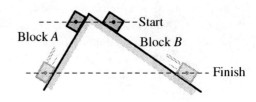

1. From the starting height to the final height, the work done on Block A by the normal force from the ramp

a) is *zero*.
b) is *negative*.
c) is *positive*.
d) *could be positive or negative* depending on the choice of coordinate systems.

Explain.

2. From the starting height to the final height, the work done on Block A by the gravitational force (the weight)

a) is *zero*.
b) is *negative*.
c) is *positive*.
d) *could be positive or negative* depending on the choice of coordinate systems.

Explain.

3. From the starting height to the final height, the work done on Block A by the gravitational force is

a) *greater than* the work done on Block B by the gravitational force.
b) *less than* the work done on Block B by the gravitational force.
c) *equal to* the work done on Block B by the gravitational force.

Explain.

4. The speed of Block A as it crosses the lower dashed line is

a) *greater than* the speed of Block B as it crosses the lower dashed line.
b) *less than* the speed of Block B as it crosses the lower dashed line.
c) *equal to* the speed of Block B as it crosses the lower dashed line.

Explain.

5. Block A will cross the lower dashed line

a) *before* Block B.
b) *after* Block B.
c) *at the same time* as Block B.

Explain.

NT6B-QRT19: Block on Ramp with Friction—Work

A block is pushed at constant speed up a ramp from point A to point B. The direction of the force on the block by the hand is horizontal. There is friction between the block and the ramp. The distance between points A and B is 1 meter.

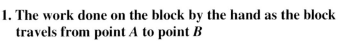

1. The work done on the block by the hand as the block travels from point A to point B

a) is *zero*.
b) is *negative*.
c) is *positive*.
d) *could be positive or negative* depending on the choice of coordinate systems.

Explain.

2. The work done on the block by the normal force from the ramp as the block travels from point A to point B

a) is *zero*.
b) is *negative*.
c) is *positive*.
d) *could be positive or negative* depending on the choice of coordinate systems.

Explain.

3. The work done on the block by the friction force from the ramp as the block travels from point A to point B

a) is *zero*.
b) is *negative*.
c) is *positive*.
d) *could be positive or negative* depending on the choice of coordinate systems.

Explain.

4. The work done on the block by the gravitational force of the earth as the block travels from point A to point B

a) is *zero*.
b) is *negative*.
c) is *positive*.
d) *could be positive or negative* depending on the choice of coordinate systems.

Explain.

NT6B-QRT20: BLOCK ON RAMP WITH FRICTION–WORK AND ENERGY

A block is pushed at constant speed up a ramp from point A to point B. The direction of the force on the block by the hand is horizontal. There is friction between the block and the ramp. The distance between points A and B is 1 meter.

1. The kinetic energy of the block at point B

a) is *greater than* the kinetic energy of the block at point A.

b) is *less than* the kinetic energy of the block at point A.

c) is *equal to* the kinetic energy of the block at point A.

d) *cannot be compared* to the kinetic energy of the block at point A unless we know the height difference between A and B.

Explain.

2. The net work done on the block as it travels from point A to point B

a) is *zero*.

b) is *negative*.

c) is *positive*.

d) *could be positive or negative* depending on the choice of coordinate systems.

Explain.

3. The work done on the block by the hand as the block travels from point A to point B

a) is *equal to* 1 meter times the magnitude of the force exerted on the block by the hand.

b) is *greater than* 1 meter times the magnitude of the force exerted on the block by the hand.

c) is *less than* 1 meter times the magnitude of the force exerted on the block by the hand but not zero.

d) is *zero*.

e) *cannot be compared* to the magnitude of the force exerted on the block by the hand based on the information given.

Explain.

NT6C-WWT21: FORCE VS. POSITION GRAPH I—WORK DONE ON BOX

A 10-kg box initially at rest is pushed a distance of 8 m along a smooth horizontal floor. A graph of the applied horizontal force on the block as a function of displacement is shown below.

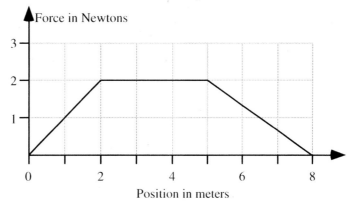

A student calculates that the work done by the applied force during the first 2 meters was 4 J and that the work done during the following 3 meters was 6 J.

What, if anything, is wrong with this calculation? If something is wrong, identify it and explain how to correct it. If this calculation is correct, explain why.

NT6C-CT22: FORCE VS. POSITION GRAPH II—WORK DONE ON BOX

A 10-kg box initially at rest is pushed a distance of 8 m along a smooth horizontal floor. A graph of the applied horizontal force on the block as a function of displacement is shown below.

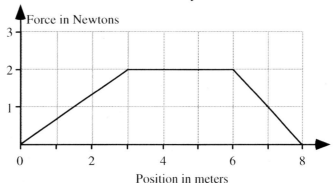

Will the work done by the applied force over the first 3 meters be _greater than, less than,_ or _equal to_ the work done by the applied force during the subsequent 3 meters?

Explain.

NT6C-RT23: Force vs. Position Graph I—Work Done on Box

A 10-kg box initially at rest is pushed a distance of 8 m along a smooth horizontal floor. A graph of the applied horizontal force on the block as a function of displacement is shown below.

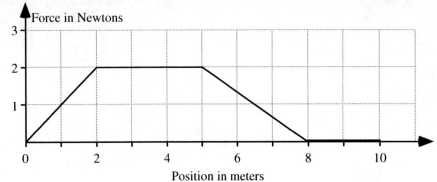

Rank the work done on the box during each 2-meter interval indicated below.

A. 0 to 2 m   B. 2 to 4 m   C. 4 to 6 m

D. 6 to 8 m   E. 8 to 10 m

        Greatest   1 _____ 2 _____ 3 _____ 4 _____ 5 _____ Least

OR, The work done on the box during each of the intervals is the same but not zero.   ___

OR, The work done on the box during each of the intervals is zero.   ___

OR, We cannot determine the ranking of the work done during the intervals.   ___

Please explain your reasoning.

NT6C-RT24: FORCE VS. POSITION GRAPH III—ENERGY TRANSFER

The graph below shows the force that an employee exerts on a cart loaded with wood at a lumberyard. This force varies as a function of position. Six segments are marked in the graph.

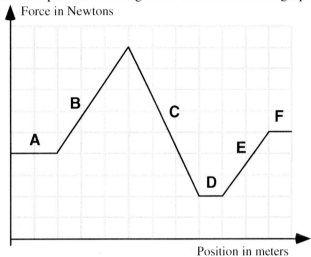

Rank these segments on the basis of the energy that the employee transfers to the cart.

Greatest   1 _____ 2 _____ 3 _____ 4 _____ 5 _____ 6 _____ Least

OR, The energy transfer is the same but not zero for all these cases.   ____

OR, The energy transfer is zero for all of these cases.   ____

OR, We cannot determine the ranking for the energy transfer here.   ____

Please explain your reasoning.

NT6C-RT25: MASS ATTACHED TO A SPRING—WORK TO STRETCH SPRING

The figures below show systems consisting of a block attached to a spring. Each block is resting on a frictionless surface. In each case, a student pulls on the block and stretches the spring to the right by the distance given in the figure. The mass of the block and force constant of the spring are also given for each case.

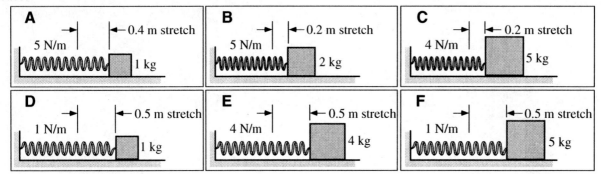

Rank these systems on the basis of the work done on the block-spring systems by the student.

Greatest   1 _____ 2 _____ 3 _____ 4 _____ 5 _____ 6 _____ Least

OR, The work done on the systems to stretch the springs is the same but not zero for these systems.   ___

OR, The work done on the systems to stretch the springs is zero for these systems.   ___

OR, We cannot determine the ranking for the work done on the systems to stretch the springs.   ___

Please explain your reasoning.

NT6D-CT26: Dropped and Thrown Rock—Kinetic Energy

Rock *A* is dropped from the top of a cliff at the same instant that an identical Rock *B* is thrown horizontally away from the cliff. Each of the following graphs describes part of the motion of the rocks. Use a coordinate system in which up is the positive vertical direction and the positive horizontal direction is away from the cliff with the origin at the point the balls were released.

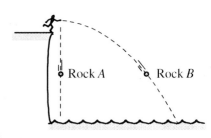

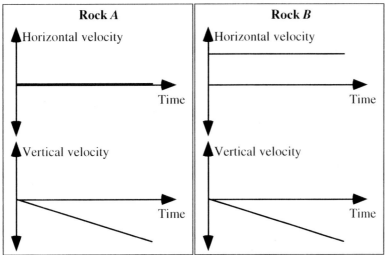

1) Is the kinetic energy of the dropped Rock *A* at the start of the motion *greater than*, *less than*, or *the same as* the kinetic energy of the thrown Rock *B*?

Explain.

2) Is the kinetic energy of the dropped Rock *A* halfway down *greater than*, *less than*, or *the same as* the kinetic energy of thrown Rock *B* halfway down?

Explain.

NT6D-CT27: Speedboats with the Same Kinetic Energy—Force

Shown are two speedboats that are racing. Both boats are moving at constant speeds on a straight stretch of the course. At this point, the speedboat on the left has more kinetic energy than the boat on the right.

KE = 180 kJ   KE = 160 kJ

Is the net force on the speedboat on the left *greater than, less than,* or *equal to* the net force on the speedboat on the right?

Explain.

NT6D-CT28: Thrown Javelins—Horizontal Force

Shown are two javelins (light spears) that have been thrown at targets. We are viewing the javelins when they are in the air about halfway to landing. Both javelins have the same mass, but they have different kinetic energies as shown. (Ignore air resistance for this task.)

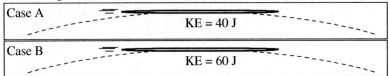

Is the horizontal force acting on the javelin in Case A *greater than, less than,* or *equal to* the horizontal force acting on the javelin in Case B?

Explain.

NT6D-CCT29: Skaters Pushing off Each Other—Force

Two skaters—a small girl and a large boy—are initially standing face-to-face but then push off each other. After they are no longer touching, the girl has more kinetic energy than the boy. Three physics students make the following contentions about the forces the boy and girl exerted on each other:

Arianna: *"I think the boy pushed harder on the girl because he is bigger, so she ended up with more kinetic energy than he did."*

Boris: *"I disagree. They pushed equally hard on each other, but the girl moved farther while they were pushing on each other, so she ended up with more kinetic energy."*

Carmen: *"I think the girl had to push harder to get the boy moving since he is bigger, but that caused her to accelerate more as she recoiled."*

Which, if any, of these three students do you agree with?

Arianna_____ Boris _____ Carmen _____ None of them_____

Explain.

NT6D-QRT30: Signs of Kinematic Quantities—Location, Velocity, & Kinetic Energy

Eight possible sign combinations for the instantaneous position, velocity, and acceleration of an object are given in the table below. Above the table is a coordinate axis that shows the origin, marked 0, and that indicates that the positive direction is to the right. The three columns on the right-hand side of the table are to describe the location of the object (either left or right of the origin), the direction of the velocity of the object (either toward or away from the origin), and what is happening to the kinetic energy of the object (either increasing or decreasing) at the given instant. The appropriate descriptions for the first case are shown.

Complete the rest of the table for the location, direction of the velocity, and kinetic energy of the object.

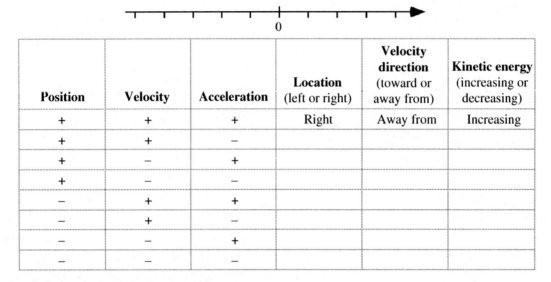

Position	Velocity	Acceleration	Location (left or right)	Velocity direction (toward or away from)	Kinetic energy (increasing or decreasing)
+	+	+	Right	Away from	Increasing
+	+	−			
+	−	+			
+	−	−			
−	+	+			
−	+	−			
−	−	+			
−	−	−			

NT6D-CRT31: Box Moving Up I—Kinetic Energy & Velocity Graphs

A 100-N box is initially moving upward at 4 m/s. A woman is applying a constant upward vertical force of 120 N to the box with her hand as shown.

Sketch a graph of the velocity and kinetic energy of the box as a function of time.

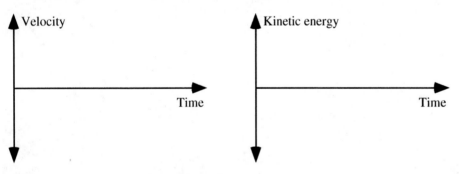

Explain.

NT6D-QRT32: SIGNS OF KINEMATIC QUANTITIES —LOCATION, NET FORCE, & KINETIC ENERGY

Eight possible sign combinations for the instantaneous position, velocity, and acceleration of an object are given in the table below. Above the table is a coordinate axis that shows the origin, marked 0, and that indicates that the positive direction is to the right. The three columns on the right-hand side of the table are to describe the location of the object (either left or right of the origin), the direction of the net force acting on the object (again left or right), and what is happening to the kinetic energy of the object (either increasing or decreasing) at the given instant. The appropriate descriptions for the first case are shown.

Complete the rest of the table for the location of, net force on, and kinetic energy of the object.

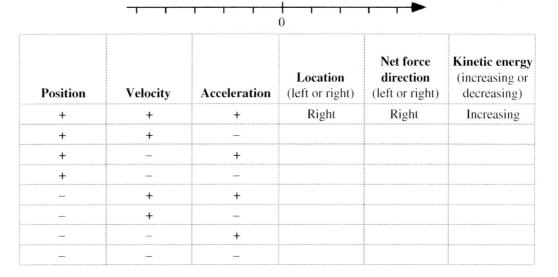

Position	Velocity	Acceleration	Location (left or right)	Net force direction (left or right)	Kinetic energy (increasing or decreasing)
+	+	+	Right	Right	Increasing
+	+	−			
+	−	+			
+	−	−			
−	+	+			
−	+	−			
−	−	+			
−	−	−			

NT6D-CRT33: BOX MOVING UP II—KINETIC ENERGY & VELOCITY GRAPHS

A 100-N box is initially moving upward at 4 m/s. A woman is applying a vertical force of 80 N with her hand to the box as shown.

Sketch a graph of the velocity and kinetic energy of the box as a function of time for the first 2 seconds.

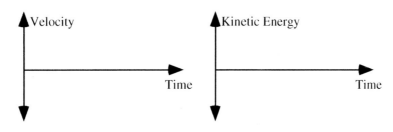

Explain.

NT6E-RT34: POTENTIAL ENERGY VS. POSITION GRAPH I—FORCE MAGNITUDE

An object is located in a region of space where there may be a force acting on it. Shown is a graph of the potential energy of the system versus the position in the *x*-direction of the object for this region. Various points in this region are labeled.

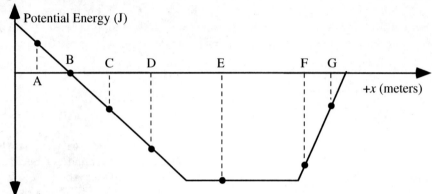

Rank the magnitude of the *x*-component of the force on the object at the labeled points.

Greatest   1 _____ 2 _____ 3 _____ 4 _____ 5 _____ 6 _____ 7 _____   Least

OR, The magnitude of the force on the object is the same (but not zero) for all of these points.   ____

OR, The magnitude of the force on the object is zero for all of these points.   ____

OR, The ranking of the magnitude of the force on the object cannot be determined.   ____

Explain your reasoning.

NT6E-RT35: POTENTIAL ENERGY VS. POSITION GRAPH II—FORCE MAGNITUDE

A spaceship is located in a region of space where there may be a force acting on it. Shown below is a graph of the potential energy as a function of the spaceship's position in the *x*-direction along its path. The points labeled in the graph represent specific locations along the path.

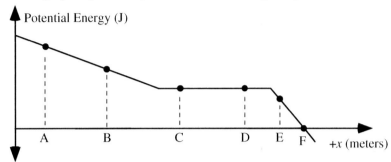

Rank the strength (magnitude) of the force on the spaceship at the labeled points.

Greatest   1 _____ 2 _____ 3 _____ 4 _____ 5 _____ 6 _____ Least

OR, The magnitude of the force on the spaceship is the same but not zero for all these locations. ____

OR, The magnitude of the force on the spaceship is zero for all these locations.   ____

OR, The ranking of the magnitude of the force on the spaceship cannot be determined.   ____

Explain your reasoning.

NT6E-RT36: Stacked Blocks Sets V—Gravitational Potential Energy

Shown below are stacks of various blocks. The masses are given in the diagram in terms of M, the mass of the smallest block. Each block has the same height and has its center of mass in the center of the block. Consider the zero point for the gravitational potential energy to be at the height of the center of mass of the lowest block.

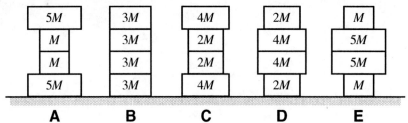

Rank the gravitational potential energy of each stack.

Greatest   1 _____ 2 _____ 3 _____ 4 _____ 5 _____   Least

OR, The gravitational potential energy of each stack is the same.   ___

OR, The ranking for the gravitational potential energy of the stacks cannot be determined.   ___

Explain your reasoning.

NT6E-RT37: TWO BLOCKS, RAMP, & SPRING—SPRING POTENTIAL ENERGY

In each figure below, a system consisting of two blocks that are tied together is attached to a spring. The systems are all at rest, and all of the springs are stretched by the same amount. The blocks on the horizontal surface are all identical, but the masses of the blocks on the inclined surfaces and the spring constants of the springs vary as shown in the figures. There may be friction between the blocks and the surfaces.

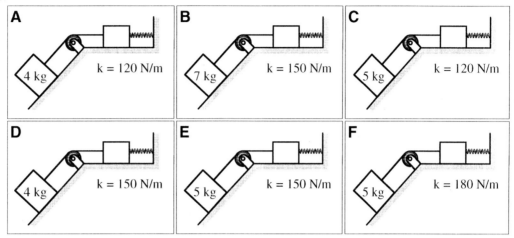

Rank these cases on the basis of the spring potential energy.

Greatest   1 _____ 2 _____ 3 _____ 4 _____ 5 _____ 6 _____ Least

OR, The spring potential energy is the same for all these systems, but it is not zero.   ____

OR, The spring potential energy is zero for all these systems.   ____

OR, We cannot determine the ranking for the spring potential energy.   ____

Please explain your reasoning.

NT6E-RT38: BLOCKS ON STRETCHED HORIZONTAL SPRINGS—POTENTIAL ENERGY

In each figure below, a system consisting of a block attached to the end of a spring is resting on a frictionless surface. The blocks are moved to the right so that the springs are stretched by the distance given in the figure. The mass and spring constant are also given for each system.

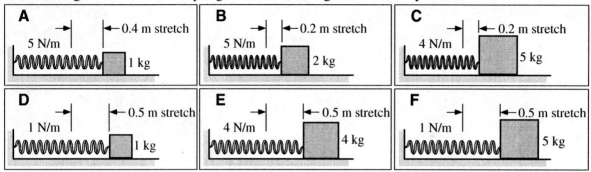

Rank these systems on the basis of the potential energy of the stretched spring-block systems.

Greatest   1 _____ 2 _____ 3 _____ 4 _____ 5 _____ 6 _____ Least

OR, The potential energy of the spring-block systems will be the same but not zero.   ___

OR, The potential energy of the spring-block systems will be zero for all these systems.   ___

OR, We cannot determine the ranking for the potential energy of the spring-block systems. ___

Please explain your reasoning.

NT6F-RT39: ARROWS SHOT FROM BUILDINGS—FINAL SPEED

In each case below, an arrow has been shot from the top of a building either up at a 45° angle, straight out horizontally, or down at a 45° angle. All arrows are identical and are shot at the same speed, and the heights of the buildings and the direction the arrows are shot are given. Ignore air resistance.

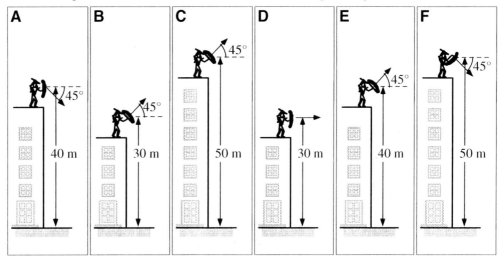

Rank these arrows on the basis of their speeds just before they hit the ground below.

Greatest   1 _____ 2 _____ 3 _____ 4 _____ 5 _____ 6 _____ Least

OR, The speed is the same for all these arrows but not zero.   ____

OR, The speed is zero for all these arrows.   ____

OR, We cannot determine the ranking for the speeds of the arrows.   ____

Please explain your reasoning.

NT6F-RT40: Toboggans Going Down Slippery Hills—Speed at Bottom

In each case below, a toboggan starts from rest and slides without friction down a snowy hill. The toboggans are all identical, and the starting heights (vertical distance above the flat bottom of the incline) and angles of the hills are given.

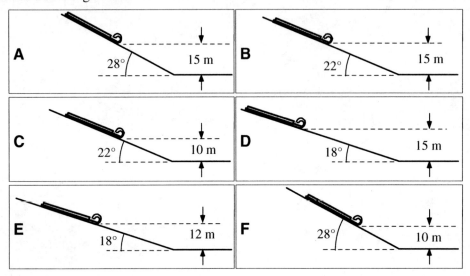

Rank these situations on the basis of the speed of the toboggan at the bottom of the incline.

Greatest   1 _____ 2 _____ 3 _____ 4 _____ 5 _____ 6 _____ Least

OR, The speed is the same for all these toboggans but it is not zero.   ____

OR, The speed is zero for all these toboggans.   ____

OR, We cannot determine the ranking for the speed of these toboggans.   ____

Please explain your reasoning.

NT6F–CT41: SKATEBOARDERS ON A HILL–TIME, SPEED, KINETIC ENERGY, AND WORK

Starting from rest, Angel and Britney skateboard down a hill as shown. Angel rides down the steep side while Britney rides down the shallow side. Angel has more mass than Britney. Assume that friction and air resistance are negligible.

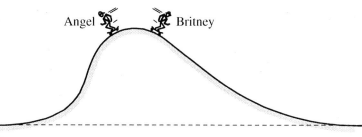

a) Is the speed at the bottom of the hill *greater for Angel, greater for Britney,* or *the same for both skateboarders*?

Explain.

b) Is the time it takes to get to the bottom of the hill *greater for Angel, greater for Britney,* or *the same for both skateboarders?*

Explain.

c) Is the work done by the gravitational force on the skateboarder *greater for Angel, greater for Britney,* or *the same for both skateboarders?*

Explain.

d) Is the work done by the normal force on the skateboarder *greater for Angel, greater for Britney,* or *the same for both skateboarders?*

Explain.

e) Is the kinetic energy at the bottom of the hill *greater for Angel, greater for Britney,* or *the same for both skateboarders*?

Explain.

NT6F-RT42: HORIZONTAL MOVING BLOCKS CONNECTED TO SPRINGS—KINETIC ENERGY

In each case below a block has been pushed against a spring on a frictionless horizontal surface, compressing the spring. The blocks were then released from rest, and at the instant shown they are just about to lose contact with the end of the spring that had been compressed. The mass of the blocks, the force constant of the springs, and the amount the springs were initially compressed are given for each system.

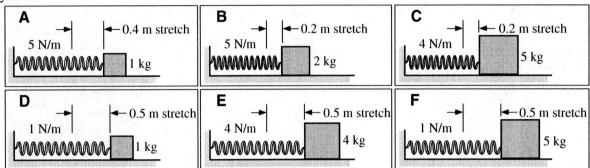

Rank these cases on the basis of the kinetic energy of the blocks at the instant shown.

Greatest   1 _____ 2 _____ 3 _____ 4 _____ 5 _____ 6 _____ Least

OR, The kinetic energy of the blocks will be the same for all these blocks.   ___

OR, We cannot determine the ranking for the kinetic energy of these blocks.   ___

Please explain your reasoning.

NT6F-CT43: BLOCK WITH COMPRESSED SPRING ON A RAMP—HEIGHT UP RAMP

Two blocks are placed on a frictionless ramp and held against a spring that is compressed one-half meter. (The mass of the blocks and force constant of the springs are also given for each system.) The blocks are then released from rest, and the compressed spring causes the blocks to accelerate up the ramp while in contact with the blocks. At the instant shown, the blocks are just about to lose contact with the end of the spring.

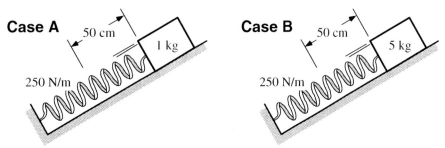

Is the height that the block travels up the ramp *greater in Case A, greater in Case B,* **or** *the same in both cases?*

Explain.

NT6F-WBT44: MATHEMATICAL RELATION—PHYSICAL SITUATION

The equation below results from the application of a physical principle to a physical system:

$$(0.5)(12 \text{ kg})(4 \text{ m/s})^2 + (12 \text{ kg})(9.8 \text{ m/s}^2)(5 \text{ m}) = (12 \text{ kg})(9.8 \text{ m/s}^2)(3.5 \text{ m}) + (0.5)(k)(1.4 \text{ m})^2$$

Draw a physical situation that would result in this equation. Explain how your drawing is consistent with the equation.

NT6F-CCT45: Block with Compressed Spring on a Ramp—Height Up Ramp

Two blocks are placed on a frictionless ramp and held against a spring that is compressed one-half meter. (The mass of the blocks and force constant of the springs are also given for each system.) The blocks are then released from rest, and the compressed springs cause the blocks to accelerate up the ramp while the springs are in contact with the blocks. At the instant shown, the blocks are just about to lose contact with the end of the springs. Three students are discussing how far the blocks will slide up the ramps.

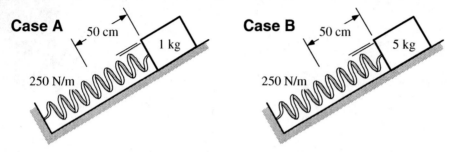

Andy: *"I think they will both travel the same distance up the inclines. The kinetic energy at the point shown in the diagram is equal to the initial potential energy stored in the compressed spring. This is the same for both cases since they both are compressed the same distance and have the same spring constants. The kinetic energy at the point shown is equal to the gravitational potential energy at the top. Since both the kinetic energy and the gravitational potential energy depend on the mass, the mass cancels out, leaving the same heights for each case."*

Badu: *"Since they both have the same energies when they are initially compressing the springs, they have to have the same energy at the top when they stop. So the lighter mass has to go higher."*

Colin: *"I think the block in Case B will go higher since it has more mass and its momentum should be larger at the point shown since they both have the same initial potential energy."*

Which, if any, of these three students do you agree with?

Andy_____ Badu _____ Colin _____ None of them_____

Explain.

NT6F-BCT46: BLOCK PUSHED ON SMOOTH RAMP—BASIC ENERGY BAR CHART

A block is pushed so that it moves up a smooth (frictionless) ramp at constant speed from A to B.

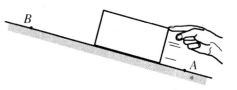

Complete the qualitative energy bar chart below for the earth-block system as the block moves from A to B. Put the zero point for the gravitational potential energy at A.

Initial system energy			During	Final system energy		
KE	PE$_{grav}$	PE$_{spring}$	W$_{ext}$	KE	PE$_{grav}$	PE$_{spring}$

Bar chart key	
KE	Kinetic energy
PE$_{grav}$	Gravitational potential energy
PE$_{spring}$	Spring potential energy
W$_{ext}$	Work done by external forces

Use g = 10 m/s² for simplicity

Explain.

NT6F-BCT47: Block Pushed on Rough Ramp—Basic Energy Bar Chart

A block is pushed so that it moves up a rough ramp at constant speed from *A* to *B*.

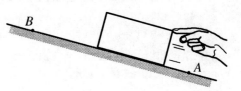

Complete the qualitative energy bar chart below for the earth-block system before and after the block has moved from *A* to *B*. Put the zero point for the gravitational potential energy at *A*.

Initial system energy			During	Final system energy		
KE	PE$_{grav}$	PE$_{spring}$	W$_{ext}$	KE	PE$_{grav}$	PE$_{spring}$

Bar chart key	
KE	Kinetic energy
PE$_{grav}$	Gravitational potential energy
PE$_{spring}$	Spring potential energy
W$_{ext}$	Work done by external forces

Use g = 10 m/s²
for simplicity

Explain.

NT6F-BCT48: Box Pulled on Smooth Surface—Basic Energy Bar Chart

A 100-N box is initially at rest at point *A* on a smooth (frictionless) horizontal surface. A student applies a horizontal force of 80 N to the right on the box as shown.

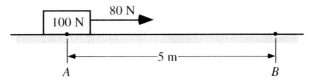

Complete the qualitative energy bar chart below for the earth-box system before and after the box has moved a horizontal distance of 5.0 m. Put the zero point for the gravitational potential energy at the surface.

Initial system energy			During	Final system energy				Bar chart key	
KE	PE$_{grav}$	PE$_{spring}$	W$_{ext}$	KE	PE$_{grav}$	PE$_{spring}$			
								KE	Kinetic energy
								PE$_{grav}$	Gravitational potential energy
							0	PE$_{spring}$	Spring potential energy
								W$_{ext}$	Work done by external forces

Use g = 10 m/s² for simplicity

Explain.

NT6F-BCT49: BOX PULLED ON ROUGH SURFACE—BASIC ENERGY BAR CHART

A 100-N box is initially at rest on a rough horizontal surface where the coefficient of static friction is 0.6 and the coefficient of kinetic friction is 0.4. A student applies a horizontal force of 80 N to the right on the box as shown. The box starts at rest at point A.

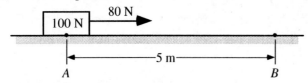

Complete the qualitative energy bar chart below for the earth-box system before and after the box has moved a horizontal distance of 5.0 m. Put the zero point for the gravitational potential energy at the surface.

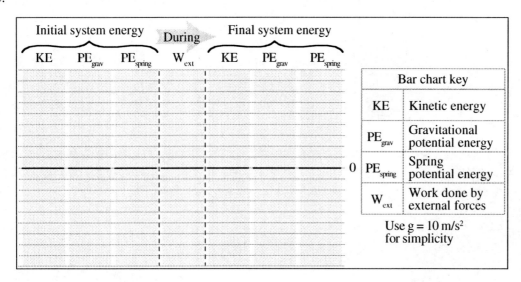

Explain.

NT6F-BCT50: LIFTED BOX MOVING UPWARD I—BASIC ENERGY BAR CHART

A 100-N box is initially 0.40 m above the surface of a table and is moving upward with a kinetic energy of 80 J. A man is applying a constant upward vertical force of 80 N with his hand to the box as shown.

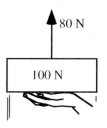

Complete the qualitative energy bar chart below for the earth-box system before and after the box has moved upward a distance of 1.0 m. Put the zero point for the gravitational potential energy at the surface of the table.

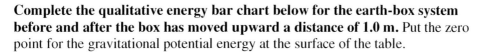

Explain.

NT6F-BCT51: SKATEBOARDER LAUNCHED BY A SPRING—BASIC ENERGY BAR CHART I

A circus performer on a skateboard is launched by a spring initially compressed a distance Δx as shown at right. His speed on the horizontal portion of the ramp is v, and he rises to a height H after he leaves the ramp. Ignore any effects due to friction.

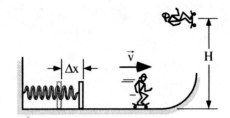

Draw a qualitative energy bar chart for the earth-skateboarder-spring system as the skateboarder goes from the compressed spring position at rest to where he moves free of the spring on the horizontal surface. Put the zero point for the gravitational potential energy at the height of the skateboarder before launching.

Initial system energy			During	Final system energy				Bar chart key	
KE	PE$_{grav}$	PE$_{spring}$	W$_{ext}$	KE	PE$_{grav}$	PE$_{spring}$		KE	Kinetic energy
								PE$_{grav}$	Gravitational potential energy
							0	PE$_{spring}$	Spring potential energy
								W$_{ext}$	Work done by external forces

Use $g = 10$ m/s^2 for simplicity

Explain.

NT6F-BCT52: SKATEBOARDER LAUNCHED BY A SPRING—BASIC ENERGY BAR CHART II

A circus performer on a skateboard is launched by a spring initially compressed a distance Δx as shown at right. His speed on the horizontal portion of the ramp is v, and he rises to a height H after he leaves the ramp. Ignore any effects due to friction.

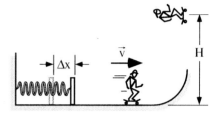

Draw a qualitative energy bar chart below for the earth-skateboarder-spring system as he goes from the compressed spring position at rest to when he reaches the height H. Put the zero point for the gravitational potential energy at the initial height of the skateboarder before launching.

Initial system energy			During	Final system energy				Bar chart key	
KE	PE$_{grav}$	PE$_{spring}$	W$_{ext}$	KE	PE$_{grav}$	PE$_{spring}$			
								KE	Kinetic energy
								PE$_{grav}$	Gravitational potential energy
							0	PE$_{spring}$	Spring potential energy
								W$_{ext}$	Work done by external forces
								Use g = 10 m/s^2 for simplicity	

Explain.

NT6F-BCT53: Box Moving Upward I—Energy Bar Chart for the Earth-Box System

A 100-N box is initially moving upward at 4 m/s. A student is applying a vertical force of 80 N to the box as shown.

Complete the energy bar chart below for the earth-box system as the box moves upward a distance of 1 m with the zero point for the gravitational potential energy set at the initial height.

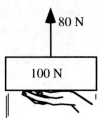

Bar chart key	
KE	Kinetic energy
E_{int}	Internal energy
PE_{grav}	Gravitational potential energy
PE_{spring}	Spring potential energy
W_{ext}	Work done by external forces
Q	Heat added to system

Use $g = 10$ m/s^2 for simplicity

Initial system energy: KE, E_{int}, PE_{grav}, PE_{spring} — During: W_{ext}, Q — Final system energy: KE, E_{int}, PE_{grav}, PE_{spring}

Explain.

NT6F-BCT54: Box Moving Upward II—Energy Bar Chart for the Earth-Box System

A 100-N box is initially moving upward at 4 m/s. A student is applying a vertical force of 80 N to the box as shown.

Complete the energy bar chart below for the earth-box system as the box moves upward a distance of 1 m with the zero point for the gravitational potential energy set at the final height.

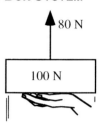

	Initial system energy			During		Final system energy				Bar chart key	
KE	E_int	PE_grav	PE_spring	W_ext	Q	KE	E_int	PE_grav	PE_spring		

Initial system energy — KE, E_{int}, PE_{grav}, PE_{spring}

During — W_{ext}, Q

Final system energy — KE, E_{int}, PE_{grav}, PE_{spring}

Bar chart key	
KE	Kinetic energy
E_{int}	Internal energy
PE_{grav}	Gravitational potential energy
PE_{spring}	Spring potential energy
W_{ext}	Work done by external forces
Q	Heat added to system

Use $g = 10 \text{ m/s}^2$ for simplicity

Explain.

NT6F-BCT55: BOX MOVING UPWARD III—ENERGY BAR CHART FOR THE BOX

A 100-N box is initially moving upward at 4 m/s. A student is applying a vertical force of 80 N to the box as shown.

Complete the energy bar chart below using only the box as a system as the box moves upward a distance of 1 m.

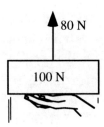

Initial system energy				During		Final system energy				Bar chart key	
KE	E_{int}	PE_{grav}	PE_{spring}	W_{ext}	Q	KE	E_{int}	PE_{grav}	PE_{spring}	KE	Kinetic energy
										E_{int}	Internal energy
										PE_{grav}	Gravitational potential energy
										PE_{spring}	Spring potential energy
										W_{ext}	Work done by external forces
										Q	Heat added to system

Use $g = 10$ m/s^2 for simplicity

Explain.

NT6F-BCT56: Box Moving Upward IV—Energy Bar Chart for the Earth–Box System

A 100-N box is initially moving upward at 4 m/s. A student is applying a vertical force of 120 N to the box as shown.

Complete the energy bar chart below for the earth-box system as the box moves upward a distance of 1 m with the zero point for the gravitational potential energy set at the initial height.

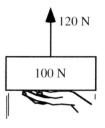

Initial system energy	During	Final system energy	Bar chart key	
KE E_{int} PE$_{grav}$ PE$_{spring}$	W_{ext} Q	KE E_{int} PE$_{grav}$ PE$_{spring}$	KE	Kinetic energy
			E_{int}	Internal energy
			PE$_{grav}$	Gravitational potential energy
			PE$_{spring}$	Spring potential energy
		0	W_{ext}	Work done by external forces
			Q	Heat added to system
			Use g = 10 m/s² for simplicity	

Explain.

NT6F-BCT57: Box Attached to a Spring I—Energy Bar Chart

A 40-N box initially held at rest is pulled a distance of 2 m upward by a constant upward vertical force of 90 N. The box is attached to a spring with a stiffness of 20 N/m that is initially not stretched or compressed.

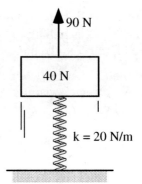

Complete the energy bar chart for the earth-box-spring system for this process.

Initial system energy				During		Final system energy				Bar chart key	
KE	E_{int}	PE_{grav}	PE_{spring}	W_{ext}	Q	KE	E_{int}	PE_{grav}	PE_{spring}	KE	Kinetic energy
										E_{int}	Internal energy
										PE_{grav}	Gravitational potential energy
										PE_{spring}	Spring potential energy
										W_{ext}	Work done by external forces
										Q	Heat added to system

Use g = 10 m/s^2 for simplicity

Explain.

NT6F-BCT57: BOX SUSPENDED BY A SPRING II—ENERGY BAR CHART

A 40-N box initially held at rest is pulled a distance of 2 m downward by a constant downward vertical force of 50 N. The box is attached to a fixed spring with a stiffness of 20 N/m that is initially not stretched or compressed.

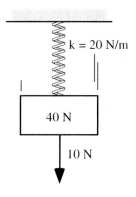

Complete the energy bar chart for the earth-box-spring system for this process.

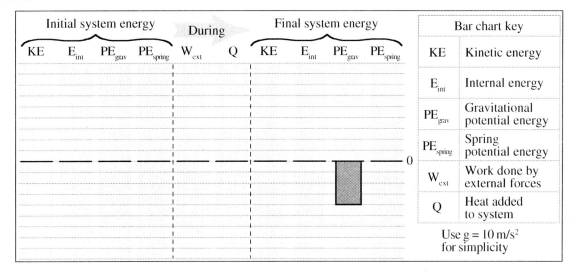

Explain.

NT6F-BCT58: BOX SUSPENDED BY A SPRING II—ENERGY BAR CHART

A 40-N box initially held at rest is pulled a distance of 2 m downward by a constant downward vertical force of 10 N. The box is attached to a spring with a stiffness of 20 N/m that is initially not stretched or compressed.

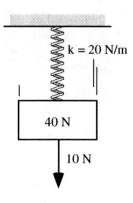

Complete the energy bar chart for the earth-box-spring system for this process.

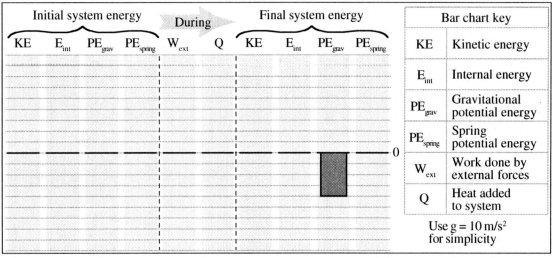

Explain.

NT6F-BCT59: Box Attached to a Spring on a Horizontal Surface—Energy Bar Chart

A 70-N box initially at rest is pulled a distance of 4 m along a rough horizontal floor by a constant horizontal force of 80 N. The box is attached to a spring with a stiffness of 20 N/m that is initially not stretched or compressed. The maximum static frictional force is 30 N, and the kinetic frictional force is 20 N between the box and surface.

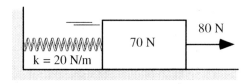

Complete the energy bar chart for the surface-box-spring system for this process.

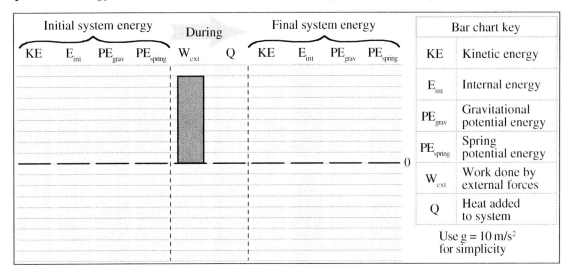

Explain.

NT6F-BCT60: BALL THROWN UPWARDS I—ENERGY BAR CHART FOR BALL & EARTH

A ball is thrown straight upwards. After reaching the top of its trajectory, it falls back to the height it was released from, but during its flight 20% of the ball's initial kinetic energy was lost due to air resistance. Assume the ball's temperature does not change during this time.

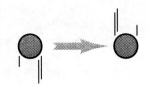

Complete the energy bar chart below for the ball-earth system for this process.

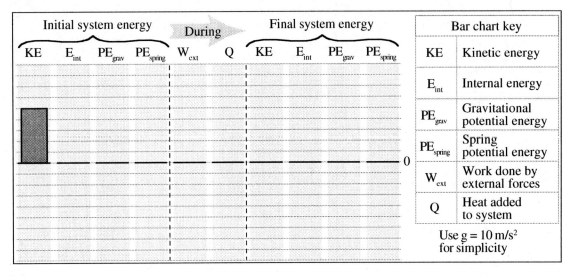

Explain.

NT6F-BCT61: BALL THROWN UPWARDS II—ENERGY BAR CHART FOR BALL & EARTH

A ball is thrown straight upwards. After reaching the top of its trajectory, it falls back to the height it was released from, but during its flight 20% of the ball's initial kinetic energy was lost due to air resistance.

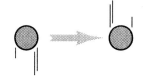

Complete the energy bar chart below for the ball-earth-air system for this process.

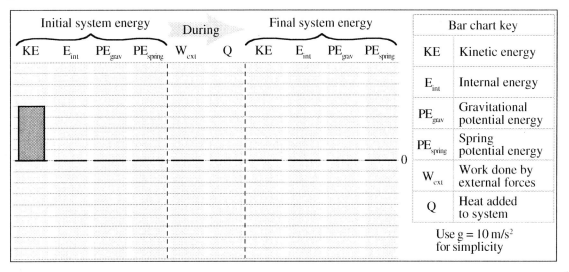

Explain.

NT6F-BCT62: Box Sliding on a Rough Ramp I—Energy Bar Chart

A 400-N box starts from rest from the top of a ramp that has a length of 1.3 m and a height of 0.5 m. The box slides to the bottom of the ramp where the gravitational potential energy is zero. The speed of the box at the bottom of the ramp is 2 m/s. There is friction on the box by the ramp.

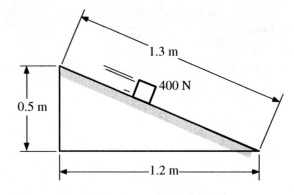

Complete the energy bar chart below for the box-earth-ramp system for this process.

Initial system energy				During		Final system energy				Bar chart key	
KE	E_{int}	PE_{grav}	PE_{spring}	W_{ext}	Q	KE	E_{int}	PE_{grav}	PE_{spring}	KE	Kinetic energy
										E_{int}	Internal energy
										PE_{grav}	Gravitational potential energy
										PE_{spring}	Spring potential energy
										W_{ext}	Work done by external forces
										Q	Heat added to system
										Use $g = 10$ m/s^2 for simplicity	

Explain.

A 40-N box is dropped from a vertical height of 3 meters above the top of an uncompressed spring that has a spring constant of 240 N/m. The box-spring system comes to rest momentarily when the spring is compressed a vertical distance of 1 meter below its initial position. (Note: The system will start oscillating after reaching this point.) Ignore air resistance.

Complete the energy bar chart below for the box-earth-spring system for this process where the zero point for the gravitational potential energy is the location of the box when the spring has been compressed 1 m (lowest point).

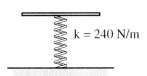

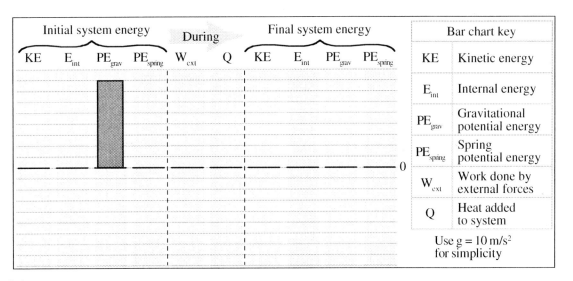

Explain.

NT6F-WWT64: BLOCK WITH COMPRESSED SPRING ON A RAMP—HEIGHT UP RAMP

Two blocks are placed on a frictionless ramp and held against a spring that is compressed one-half meter. (The mass of the blocks and force constant of the springs are given for each system.) The blocks are then released from rest, and the compressed spring causes the blocks to accelerate up the ramp while it is in contact with the blocks. At the instant shown, the blocks are just about to lose contact with the end of the spring.

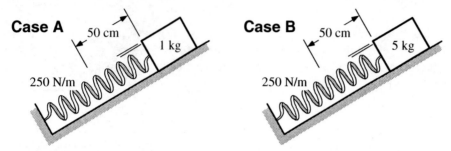

A student makes the following statement about how far the blocks will slide up the inclines:

"I think they will both travel up the same distance along the inclines. The kinetic energy at the point shown in the diagram is equal to the initial energy stored in the compressed spring. This is the same for both cases since they both are compressed the same distance and have the same spring constants. The kinetic energy at the point shown is also equal to the gravitational potential energy at the top or $mv^2/2 = mgh$. Thus, the mass cancels out, leaving the same heights for each case."

What, if anything, is wrong with this statement? If something is wrong, identify it and explain how to correct it. If this statement is correct, explain why.

NT6F-CT65: Skateboarder Launched by a Spring—Speed & Height

A circus performer on a skateboard is launched by a spring initially compressed a distance Δx as shown at right. His speed on the horizontal portion of the ramp is v, and he rises to a height H after he leaves the ramp. He then conducts a second launch with the spring initially compressed a distance $2\Delta x$.

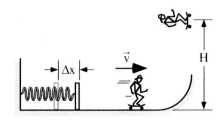

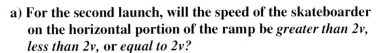

a) For the second launch, will the speed of the skateboarder on the horizontal portion of the ramp be *greater than 2v*, *less than 2v*, or *equal to 2v*?

Explain.

b) Will the height reached by the skateboarder for the second launch be *greater than 2H*, *less than 2H*, or *equal to 2H*?

Explain.

NT6F-WBT66: Energy Bar Chart I—Physical Situation

Describe a physical situation and a system to which the energy bar chart below could apply.

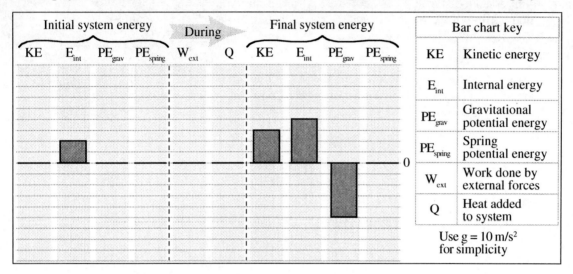

Explain.

NT6F-WBT67: ENERGY BAR CHART II—PHYSICAL SITUATION

Describe a physical situation and a system to which the energy bar chart below could apply.

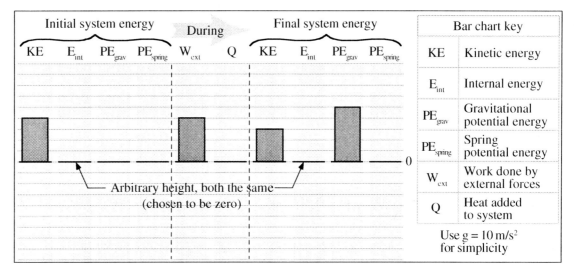

Explain.

NT6G-CT68: RACE UP A HILL—WORK AND POWER

Abbie and Bonita decide to race up a hill that is 30 meters high. Abbie takes a path that is 60 meters long while Bonita uses a path that is 100 meters long. It takes Abbie 40 seconds since her route is steep, while Bonita runs up her path in 30 seconds. They both start from rest at the same height and stop at the top. Abbie has a weight of 700 N while Bonita has a weight of 500 N.

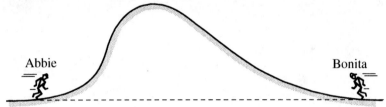

(1) Is the work that Abbie does in going up the hill *greater than*, *less than*, or *the same as* the work that Bonita does in going up the hill?

Explain.

(2) Is the power generated by Abbie in going up the hill *greater than*, *less than*, or *the same as* the power generated by Bonita in going up the hill?

Explain.

NT6G-CT69: CAR RACE—WORK AND POWER

Amanda and Bertha are in a car race. Their cars have the same mass. At one point in the race, they both change their speeds by 10 m/s in 2 seconds. Ignore air friction.

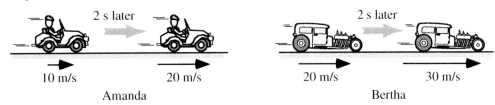

2 s later   2 s later

10 m/s   20 m/s   20 m/s   30 m/s

Amanda   Bertha

(1) Is the work that Amanda's car does while speeding up *greater than*, *less than*, or *the same as* the work that Bertha's car does while speeding up?

Explain.

(2) Is the power generated by Amanda's car while speeding up *greater than*, *less than*, or *the same as* the power generated by Bertha's car while speeding up?

Explain.

NT7 LINEAR MOMENTUM AND IMPULSE

NT7A-QRT1: POSITION AND NET FORCE SIGNS AND CHANGE IN KINETIC ENERGY—MOMENTUM

An object is moving in one dimension. The coordinate axis below shows the origin, marked 0, and the positive direction, which is to the right. The signs for the instantaneous position, the net force acting on the object, and what is happening to the kinetic energy during a 2-second interval are given in the table. The first row of the table is complete, giving the location of the object (either left or right of the origin), the direction of momentum of the object (either toward or away from the origin), and what is happening to the magnitude of the momentum of the object during the 2-second time interval (either increasing or decreasing).

Complete the rest of the table for the location, direction of momentum, and magnitude of momentum of the object.

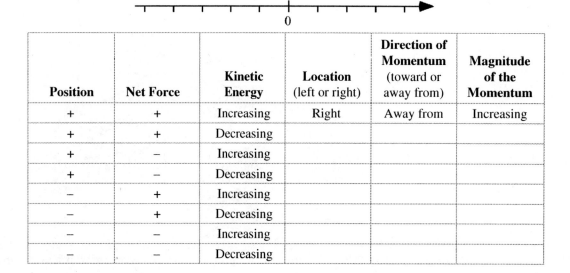

Position	Net Force	Kinetic Energy	Location (left or right)	Direction of Momentum (toward or away from)	Magnitude of the Momentum
+	+	Increasing	Right	Away from	Increasing
+	+	Decreasing			
+	–	Increasing			
+	–	Decreasing			
–	+	Increasing			
–	+	Decreasing			
–	–	Increasing			
–	–	Decreasing			

NT7A-CT2: SPEEDBOATS WITH THE SAME MOMENTUM—FORCE

Two speedboats are moving at constant speeds on a straight stretch of a racecourse. At the instant shown, Speedboat A has more momentum than Speedboat B.

Is the net force on Speedboat A *greater than, less than,* or *equal to* the net force on Speedboat B?

Explain.

nT7A-CT3: Speedboats with the Same Momentum—Kinetic Energy

Two speedboats are moving at constant speeds on a straight stretch of a racecourse. Speedboat A has more mass than Speedboat B. At the instant shown, the two speedboats have the same momentum.

Is the kinetic energy of Speedboat A *greater than, less than,* or *equal to* the kinetic energy of Speedboat B?

Explain.

nT7A-CT4: Speedboats with the Same Kinetic Energy—Momentum

Two speedboats are moving at constant speeds on a straight stretch of a racecourse. Speedboat A has more mass than Speedboat B. At the instant shown, the two speedboats have the same kinetic energy.

Is the momentum of Speedboat A *greater than, less than,* or *equal to* the momentum of Speedboat B?

Explain.

NT7A-WWT5: OBJECT CHANGING VELOCITY I—IMPULSE

A 2-kg object accelerates as a net external force is applied to it. During the 5-second interval that the force is applied, the object's velocity changes from 3 m/s east to 7 m/s west.

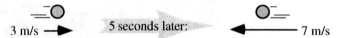

3 m/s →   5 seconds later: ← 7 m/s

A student states:

> *"The change in momentum of this object during these 5 seconds was 8 kg·m/s so the impulse applied to this object during these 5 seconds was 8/5 kg·m/s."*

What, if anything, is wrong with this statement? If something is wrong, identify and explain how to correct all errors. If this statement is correct, explain why.

NT7A-WWT6: OBJECT CHANGING VELOCITY II—IMPULSE

A 2-kg object accelerates as a net external force is applied to it. During the 5-second interval that the force is applied, the object's velocity changes from 3 m/s east to 7 m/s west.

3 m/s →   5 seconds later: ← 7 m/s

A student states:

> *"The change in velocity for this 2 kg object was 4 m/s, so the change in momentum, and also the impulse, was 8 kg·m/s."*

What, if anything, is wrong with this statement? If something is wrong, identify and explain how to correct it. If this statement is correct, explain why.

NT7A-CCT7: Object Changing Velocity III—Impulse

A 2-kg object accelerates as a net external force is applied to it. During the 5-second interval that the force is applied, the object's velocity changes from 3 m/s east to 7 m/s west.

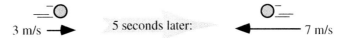

3 m/s →   5 seconds later: ← 7 m/s

Several students discussing the impulse on this object state the following:

Andre: *"The impulse is equal to the change in momentum, which is (2 kg)(3 m/s + 7 m/s) = 20 kg·m/s."*

Bela: *"But the change in velocity is 4 m/s. We multiply by the mass to get the change in momentum, and also the impulse, which is 8 kg·m/s."*

Carleton: *"The change in momentum of this object during these 5 seconds was 8 kg·m/s so the impulse applied to this object during these 5 seconds was 8/5 kg·m/s."*

Dylan: *"The impulse is the force F times the time t and since we don't know the force, we can't find the impulse for this situation."*

Which, if any, of these three students do you agree with?

Andre _____ Bela _____ Carleton _____ Dylan _____ None of them _____

Please explain your reasoning.

NT7A-WWT8: Object Changing Direction and Speed—Impulse

A student proposes the following description for the impulse on a 2-kg object that changes direction and speed as shown:

"The object goes from moving at 3 m/s in the positive x-direction to 7 m/s in the positive y-direction in 5 seconds. So the impulse given to it is 8 kg·m/s, since the impulse equals the change in momentum. The 5 seconds does not enter into the calculation of this impulse."

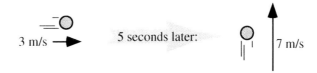

3 m/s →   5 seconds later: 7 m/s

What, if anything, is wrong with this statement? If anything is wrong, identify it and explain how to correct it. If this statement is correct, explain why.

NT7A-WWT9: OBJECT CHANGING DIRECTION AND SPEED—IMPULSE

A student proposes the following description for the impulse on a 2-kg object that changes direction and speed as shown below:

> "The impulse given to a 2 kg object that goes from moving at 3 m/s in the +x direction to 7 m/s in the +y direction in 5 seconds is (2 kg)(+7 m/s $\hat{y}$) = 14 kg·m/s $\hat{y}$ since the impulse equals the change in momentum. The 5 seconds does not enter into a calculation of this impulse."

What, if anything, is wrong with this statement? If anything is wrong, identify it and explain how to correct it. If this statement is correct, explain why.

NT7B-WWT10: FORCE VS. TIME GRAPH—IMPULSE APPLIED TO BOX

A 10-kg box, initially at rest, moves along a frictionless horizontal surface. A horizontal force to the right is applied to the box. The magnitude of the force changes as a function of time as shown.

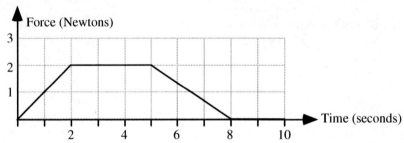

A student calculates that the impulse applied by the force during the first 2 seconds is 4 N·s and that the impulse applied during the following 3 seconds is 6 N·s.

What, if anything, is wrong with these calculations? If something is wrong, identify it and explain how to correct it. If these calculations are correct, explain why.

NT7B-WWT11: FORCE VS. TIME GRAPH—MOMENTUM VS. TIME GRAPH

A 10-kg box, initially at rest, moves along a frictionless horizontal surface. A horizontal force to the right is applied to the box. The magnitude of the force changes as a function of time as shown.

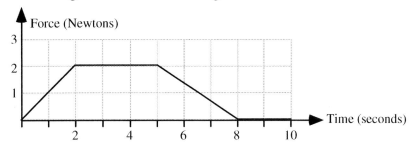

A student draws the following graph for the momentum of this 10-kg box as a function of time during this 10-second interval.

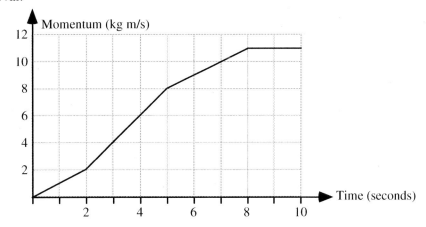

What, if anything, is wrong with this graph? If anything is wrong, identify it and explain how to correct it. If this graph is correct, explain why.

nT7B-RT12: FORCE VS. TIME GRAPH—IMPULSE APPLIED TO BOX

A 10-kg box, initially at rest, moves along a frictionless horizontal surface. A horizontal force to the right is applied to the box. The magnitude of the force changes as a function of time as shown.

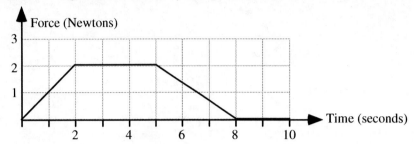

Rank the impulse applied to the box by this force during each 2-second interval indicated below.

A. 0 to 2 s   **B. 2 to 4 s**   **C. 4 to 6 s**   **D. 6 to 8 s**   **E. 8 to 10 s**

Greatest   1 _____ 2 _____ 3 _____ 4 _____ 5 _____ Least

OR, The impulse applied to the box during each of the intervals is the same but not zero.   ___

OR, The impulse applied to the box during each of the intervals is zero.   ___

OR, We cannot determine the ranking of the impulse applied to the box during the intervals. ___

Please explain your reasoning.

NT7C-WWT13: COLLIDING OBJECTS STICKING TOGETHER—MOMENTUM OF SYSTEM

An object with a mass of 4 kg moving at 5 m/s in the +y-direction collides and sticks to a second object with a mass of 6 kg moving at 3 m/s in the +x-direction as shown.

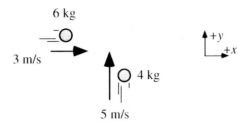

A student states:

"The total momentum of the system of both objects before the collision is 38 kg·m/s since the momentum of the 4-kg mass is 20 kg·m/s in the +y-direction and the momentum of the 6-kg mass is 18 kg·m/s in the +x-direction. The total momentum of the system after the collision is also 38 kg·m/s since momentum is conserved in collisions."

What, if anything, is wrong with this statement? If something is wrong, identify it and explain how to correct it. If this statement is correct, explain why.

NT7C-WWT14: COLLIDING OBJECTS STICKING TOGETHER—SYSTEM CENTER OF MASS VELOCITY

An object with a mass of 4 kg moving at 6 m/s in the +y-direction collides and sticks to a second object with a mass of 2 kg moving at 4 m/s in the +x-direction as shown. A student states:

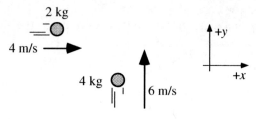

"The velocity of the center of mass of this system is the same before and after the collision. It is equal to the total momentum divided by the total mass of the system. The total momentum of the system of both objects before the collision is 32 kg·m/s since the momentum of 4-kg mass is 24 kg·m/s in the +y-direction and 2-kg mass is 8 kg·m/s in the +x-direction. Thus the velocity of the center of mass of this system is 32 kg·m/s divided by 6 kg or 5.33 m/s."

What, if anything, is wrong with this statement? If something is wrong, identify it and explain how to correct it. If this statement is correct, explain why.

NT7C-CT15: TWO BOXES ON A FRICTIONLESS SURFACE—SPEED

Two boxes are tied together by a string and are sitting at rest in the middle of a large frictionless surface. Between the two boxes is a massless compressed spring. The string tying the two boxes together is cut suddenly and the spring expands, pushing the boxes apart. The box on the left has four times the mass of the box on the right.

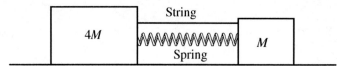

At the instant (after the string is cut) that the boxes lose contact with the spring, will the speed of the box on the left be *greater than*, *less than*, or *equal to* the speed of the box on the right?

Explain.

NT7C-WWT16: Two Skaters Pushing off Each Other—Force

Two skaters, a large girl and a small boy, are initially standing face-to-face but then push off each other. The boy ends up with more kinetic energy than the girl. A physics student who is watching makes the following contention about the forces that the boy and girl exerted on each other:

"Since the boy has more kinetic energy he also has more momentum, so the girl had to have pushed harder on him than he pushed on her."

What, if anything, is wrong with this contention? If something is wrong, identify all problems and explain how to correct them. If this contention is correct, explain why.

NT7D-WWT17: Ball Hitting a Wall—Momentum

A student observing a rubber ball hitting a wall and rebounding states:

"The change in momentum for the ball is equal and opposite to the change in momentum for the wall, because in this situation momentum has to be conserved."

What, if anything, is wrong with this statement? If something is wrong, identify it, and explain how to correct it. If this statement is correct, explain why.

NT7D-RT18: COLLIDING CARTS STICKING TOGETHER—FINAL SPEED

In each of the six figures below, two carts traveling in opposite directions are about to collide. The carts are all identical in size and shape, but they carry different loads and are initially traveling at different speeds. The carts stick together after the collision. There is no friction between the carts and the ground.

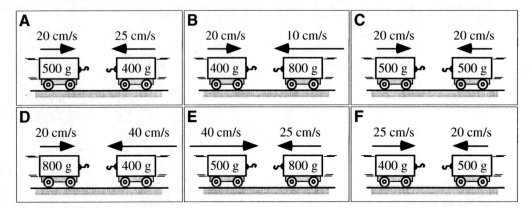

Rank these situations on the basis of the speed of the two-cart systems after the collision.

> Greatest   1 _____ 2 _____ 3 _____ 4 _____ 5 _____ 6 _____ Least

OR, The speed is the same but not zero for these two-cart systems after the collision.   ___

OR, The speed is zero for these two-cart systems after the collision.   ___

OR, We cannot determine the ranking for the speeds of these cart systems after the collision. ___

Please explain your reasoning.

NT7D-CCT19: Two Moving Carts—Result of Collision

Carts *A* and *B* are shown just before they collide. Four students discussing this situation make the following contentions:

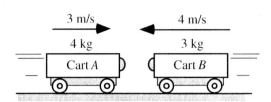

Alma: *"After the collision, the carts will stick together and move off to the left. Cart B has more speed, and its speed is going to determine which cart dominates in the collision."*

Baxter: *"I think they'll stick together and move off to the right because Cart A is heavier. It's like when a heavy truck hits a car: The truck is going to win no matter which one's going fastest, just because it's heavier."*

Callie: *"I think the speed and the mass compensate, and the carts are going to be at rest after the collision."*

Dante: *"The carts must have the same momentum after the collision as before the collision, and the only way this is going to happen is if they keep the same speeds. All the collision does is change their directions, so that Cart A will be moving to the left at 3 m/s and Cart B will be moving to the right at 4 m/s."*

Which, if any, of these four students do you agree with?

Alma_____ Baxter _____ Callie _____ Dante _____ None of them_____

Explain.

NT7D-CT20: BULLET STRIKES A WOODEN BLOCK—BLOCK & BULLET SPEED AFTER IMPACT

In Case A, a metal bullet penetrates a wooden block. In Case B, a rubber bullet with the same initial speed and mass bounces off of an identical wooden block.

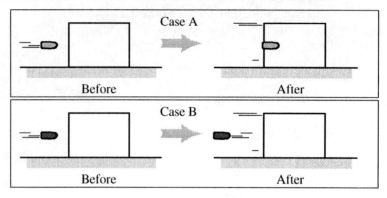

a) **Will the speed of the wooden block after the collision be *greater in Case A*, *greater in Case B*, or *the same in both cases*?**

Explain.

b) **Will the speed of the bullet in Case B after the collision be *greater than*, *less than*, or *the same as* the speed of the bullet just before the collision?**

Explain.

NT7D-QRT21: COLLIDING STEEL BALLS—CHANGE IN VELOCITY DIRECTION

Two identical steel balls, *P* and *Q*, are shown at the instant that they collide. The paths and velocities of the two balls before and after the collision are indicated by the dashed lines and arrows. The speeds of the balls are same before and after collision.

For the questions below, use the directions indicated by the arrows in the direction rosette, or use **J** for no direction, **K** for into the page, **L** for out of the page, or **M** if none of these are correct.

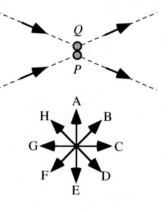

a) **Which letter best represents the direction of the change in velocity for ball *Q*?**

Explain.

b) **Which letter best represents the direction of the change in velocity for ball *P*?**

Explain.

NT7D-QRT22: COLLIDING STEEL BALLS—MOMENTUM AND IMPULSE DIRECTION

Two identical steel balls, *S* and *T*, are shown at the instant that they collide. The paths and velocities of the two balls before and after the collision are indicated by the dashed lines and arrows. The speeds of the balls are same before and after the collision.

For the questions below, use the directions indicated by the arrows in the direction rosette, or use **J** for no direction, **K** for into the page, **L** for out of the page, or **M** if none of these are correct.

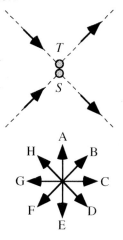

a) Which letter best represents the direction of the initial momentum of ball *T*?

Explain.

b) Which letter best represents the direction of the final momentum of ball *T*?

Explain.

c) Which letter best represents the direction of the change in momentum for ball *T*?

Explain.

(d) Which letter best represents the direction of the change in momentum for ball *S*?

Explain.

(e) Which letter best represents the direction of the impulse on ball *T*?

Explain.

f) Which letter best represents the direction of the impulse on ball *S*?

Explain.

NT7D-QRT23: System of Colliding Steel Balls—Momentum Direction of the System

Two identical steel balls, *S* and *T,* are shown at the instant that they collide. The paths and velocities of the two balls before and after the collision are indicated by the dashed lines and arrows. The speeds of the balls are same before and after the collision.

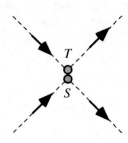

For your answers to the questions below, use the directions indicated by the arrows in the direction rosette, or use **J** for no direction, **K** for into the page, **L** for out of the page, or **M** if none of these are correct.

a) **Choose the letter that best represents the direction of the initial momentum for the system of both balls *S* and *T* before collision.**

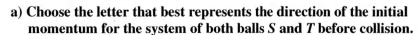

b) **Choose the letter that best represents the direction of the final momentum for the system of both balls *T* and *S* after collision.**

c) **Choose the letter that best represents the direction of the impulse during this interaction for the system of both balls *T* and *S*.**

NT7D-CCT24: COLLIDING CARTS THAT STICK TOGETHER—FINAL KINETIC ENERGY

Two identical carts traveling in opposite directions are shown just before they collide. The carts carry different loads and are initially traveling at different speeds. The carts stick together after the collision.

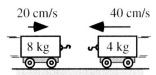

Three physics students discussing this situation make the following contentions:

Alex: *"These carts will both be at rest after the collision since the initial momentum of the system is zero, and the final momentum has to be zero also."*

Belinda: *"If that were true it would mean that they would have zero kinetic energy after the collision and that would violate conservation of energy. Since the right-hand cart has more kinetic energy, the combined carts will be moving slowly to the left after the collision."*

Chano: *"I think that after the collision the pair of carts will be traveling left at 20 cm/s. That way conservation of momentum and conservation of energy are both satisfied."*

Which, if any, of these three students do you think is correct?

Alex _____ Belinda _____ Chano _____ None of them_____

Please explain your reasoning.

NT7E-TT25: TAPERED PLATE—HORIZONTAL POSITION OF THE CENTER OF MASS

The figure below shows a plate whose height tapers from $H/2$ at its left side to double that height, H, at its right side. The plate is of uniform thickness t, is made of the same material throughout, and has a width of L, a mass M, and a mass per unit area of σ.

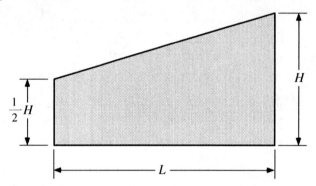

A student is trying to determine the location of the center of the mass for the beam, measured from the left end. He sets up the following integral to calculate the horizontal position of the center of the mass for the beam from the left end as

$$x_{cm} = \frac{\sigma}{M}\int_0^H h\,dx = \frac{\sigma}{M}\int_0^H \left(L/2 + \frac{H}{2L}x\right)dx$$

There is at least one thing wrong with this equation. Please identify the error(s) and explain how to correct it (them).

NT7E-TT26: TRIANGLE PLATE—HORIZONTAL POSITION OF THE CENTER OF MASS

The figure below shows a triangular plate. The plate has uniform thickness t and is made of the same material throughout. It has a height of H, a width of L, a mass M, and a mass per unit area of σ.

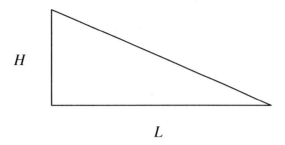

A student is trying to determine the location of the center of the mass for the plate from the left end. He sets up the following integral to calculate the horizontal (x) position of the center of the mass for the plate measured from the left end as

$$x_{cm} = \frac{\sigma}{M} \int_0^L \left(\frac{H}{L} x \right) dx$$

There is at least one thing wrong with this equation. Please identify the error(s) and explain how to correct it (them).

NT7E-TT27: CARPENTER'S SQUARE—CENTER OF MASS OF PARTS

A carpenter's square has the shape of an L with outside edges of 14 cm and 20 cm and width of 4 cm as shown below. A student is working on locating the center of mass of two parts of this square. She is going to measure it from the origin of the coordinate axis shown at right below. She divides the carpenter's square into two parts A_1 and A_2 as shown.

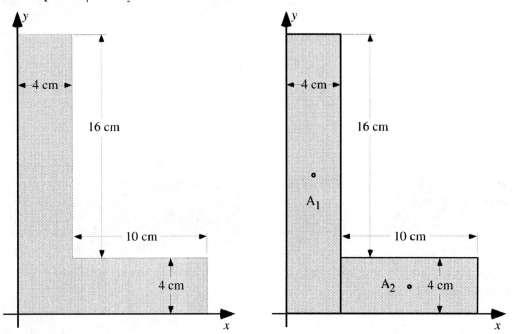

She determines the area of A_1 to be 80 cm^2 with the center at (2 cm, 10 cm) and the area of A_2 to be 40 cm^2 with the center at (5 cm, 2 cm) as shown with the dots in the drawing on the right above.

There is something wrong with this. Identify it and explain how to correct it.

NT8A-WWT1: ANGULAR VELOCITY VS. TIME GRAPH—ANGULAR ACCELERATION VS. TIME GRAPH

A student obtains a graph of an object's angular velocity versus time and then draws the graph of the angular acceleration versus time for the same time interval.

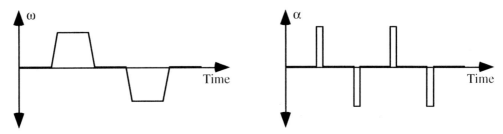

What, if anything, is wrong with this graph of the angular acceleration versus time? If something is wrong, explain how to correct it. If this graph of the angular acceleration is correct, explain why.

NT8A-CRT2: PULLEY AND WEIGHT—ANGULAR VELOCITY AND ACCELERATION GRAPHS

A weight is tied to a rope that is wrapped around a pulley. The pulley is initially rotating counterclockwise and is pulling the weight up. The tension in the rope creates a torque on the pulley that opposes this rotation.

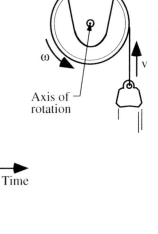

a) **On the axes below, draw a graph of the angular velocity versus time for the period from the initial instant shown until the weight comes back down to the same height.** Take the initial angular velocity as positive.

b) **Draw a graph of the angular acceleration versus time for the same time period.**

Explain.

NT8A-CRT3: ANGULAR ACCELERATION VS. TIME GRAPH—ANGULAR VELOCITY VS. TIME GRAPH

Sketch an angular velocity versus time graph given the angular acceleration graph shown for the same time interval, assuming the initial angular velocity is zero.

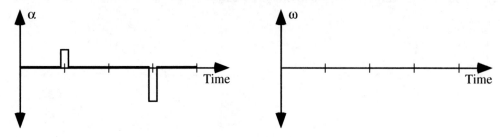

Explain.

NT8A-CRT4: ANGULAR VELOCITY VS. TIME GRAPH—ANGULAR ACCELERATION VS. TIME GRAPH

Sketch an angular acceleration versus time graph given the angular velocity versus time graph shown for the same time interval.

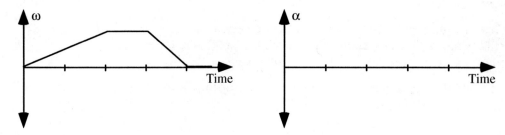

Explain.

NT8A-RT5: STOPPING JET ENGINES—ANGULAR ACCELERATION

In each figure below, the jet engine is slowing down due to the application of a constant torque. All of the engines are identical, but they start with different angular speeds and have torques of different magnitudes applied to the rotating shafts within the engines. Magnitudes of the initial angular speeds and torques are given in the figures.

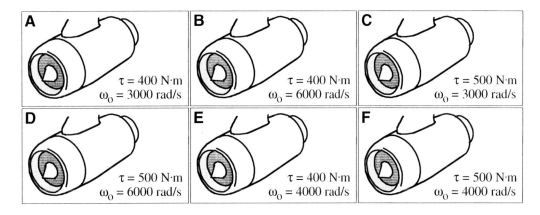

A
$\tau = 400$ N·m
$\omega_0 = 3000$ rad/s

B
$\tau = 400$ N·m
$\omega_0 = 6000$ rad/s

C
$\tau = 500$ N·m
$\omega_0 = 3000$ rad/s

D
$\tau = 500$ N·m
$\omega_0 = 6000$ rad/s

E
$\tau = 400$ N·m
$\omega_0 = 4000$ rad/s

F
$\tau = 500$ N·m
$\omega_0 = 4000$ rad/s

Rank these situations on the basis of the magnitude of the angular acceleration of the engines as they slow down.

Greatest   1 _____ 2 _____ 3 _____ 4 _____ 5 _____ 6 _____ Least

OR, The magnitude of the angular acceleration is the same but not zero for all these engines. ___

OR, The magnitude of the angular acceleration is zero for all these engines.   ____

OR, We cannot determine the ranking for the angular accelerations of these engines.   ___

Please explain your reasoning.

NT8A-CT6: ROTATING DISC—ACCELERATION AND VELOCITY

A disc with a moment of inertia of 0.2 kg·m^2 rotates at 300 revolutions per minute. It takes 40 s for the disc to reach this rotation rate starting from rest. Consider a point A on the disc that is 1 cm from the axis of rotation and another point B that is farther from this axis.

Disc rotating clockwise – top view

Ten seconds after starting from rest:

(1) Will the magnitude of the angular acceleration of point A be *greater than*, *less than*, or *equal to* the magnitude of the angular acceleration of point B?

Explain.

(2) Will the magnitude of the angular velocity of point A be *greater than*, *less than*, or *equal to* the magnitude of the angular velocity of point B?

Explain.

(3) Will the magnitude of the linear velocity of point A be *greater than*, *less than*, or *equal to* the magnitude of the linear velocity of point B?

Explain.

NT8A-CT7: ANGULAR VELOCITY VS. TIME GRAPHS—ANGULAR DISPLACEMENT

The graphs below show the angular velocity of two objects (labeled as object 1 and object 2) during the same time interval.

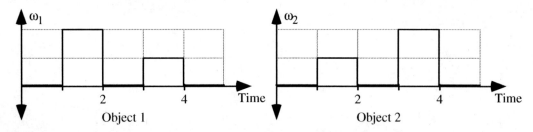

After the 5-second time interval, is the angular displacement of Object 1 *greater than*, *equal to*, or *less than* the angular displacement of Object 2?

Explain.

NT8B-RT8: RIGID 3-DIMENSIONAL POINT OBJECTS—MOMENT OF INERTIA ABOUT x-AXIS

Shown below are six small brass and aluminum spheres connected by three stiff lightweight rods to form a rigid object. The rods are joined at their centers, are mutually perpendicular, and lie along the axes of the coordinate system shown. All spheres are the same distance from the connection point of the three rods at the origin of the coordinate axis. The brass spheres are shaded in the diagram and are identical. The aluminum spheres are identical, have less mass than the brass spheres, and are unshaded in the diagram. For this problem, treat the metal spheres as point masses and ignore the mass of the connecting rods.

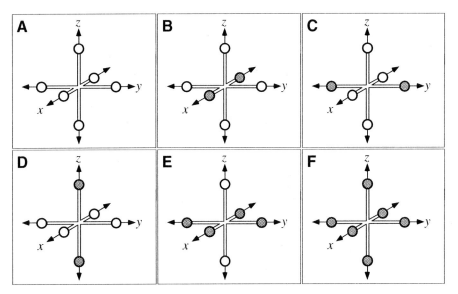

Rank these objects on the basis of the moment of inertia about the x-axis.

Greatest   1 _____ 2 _____ 3 _____ 4 _____ 5 _____ 6 _____ Least

OR, The moment of inertia about the x-axis is the same but not zero for all these objects.   ____

OR, The moment of inertia about the x-axis is zero for all these objects.   ____

OR, The ranking for the moments of inertia about the x-axis cannot be determined.   ____

Explain your reasoning.

NT8B-RT9: RIGID 3-DIMENSIONAL POINT OBJECTS—MOMENT OF INERTIA ABOUT Z-AXIS

Shown below are six small brass and aluminum spheres connected by three stiff lightweight rods to form a rigid object. The rods are joined at their centers, are mutually perpendicular, and lie along the axes of the coordinate system shown. All spheres are the same distance from the connection point of the three rods at the origin of the coordinate axis. The brass spheres are shaded in the diagram and are identical. The aluminum spheres are identical, have less mass than the brass spheres, and are unshaded in the diagram. For this problem, treat the metal spheres as point masses and ignore the mass of the connecting rods.

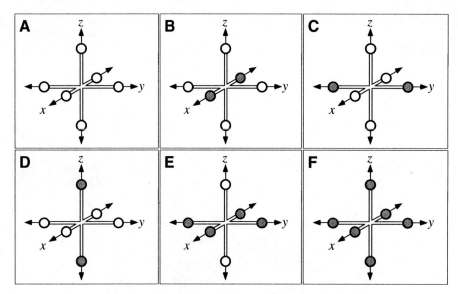

Rank these objects on the basis of the moment of inertia about the z-axis.

Greatest   1 _____ 2 _____ 3 _____ 4 _____ 5 _____ 6 _____ Least

OR, The moment of inertia about the z-axis is the same but not zero for all these objects.   ____

OR, The moment of inertia about the z-axis is zero for all these objects.   ____

OR, The ranking for the moments of inertia about the z-axis cannot be determined.   ____

Explain your reasoning.

NT8B-RT10: Flat Objects—Moment of Inertia Perpendicular to Surface

Three flat objects (circular ring, circular disc, and square loop) have the same mass M and the same outer dimension (circular objects have diameters of $2R$ and the square loop has sides of $2R$). The small circle at the center of each figure represents the axis of rotation for these objects. This axis of rotation passes through the center of mass and is perpendicular to the plane of the objects.

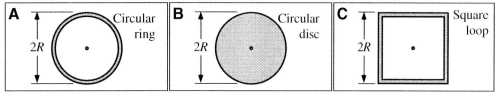

Rank the moment of inertia of these objects about this axis of rotation.

Greatest   1 _____ 2 _____ 3 _____   Least

OR, The moment of inertia of these objects is the same. ____

OR, We cannot determine the ranking for the moment of inertia of these objects. ____

Please explain your reasoning.

NT8C-QRT11: PULLEYS WITH DIFFERENT RADII—ROTATION AND TORQUE

A wheel is composed of two pulleys with different radii (labeled *a* and *b*) that are attached to one another so that they rotate together. Each pulley has a string wrapped around it with a weight hanging from it as shown. The whole system is free to rotate about a horizontal axis at the center. When the wheel is released it is found to have an angular acceleration that is directed out of the page.

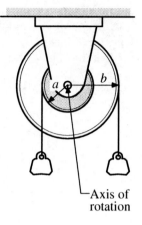

Axis of rotation

(1) Which way is the wheel rotating?

Explain.

(2) What is the direction of the net torque on the system?

How do you know?

(3) How do the masses of the two weights compare?

Explain.

NT8C-CT12: MERRY-GO-ROUNDS—NET TORQUE

Shown below are two merry-go-rounds that are identical and rotating at different constant rates. The angular velocity of the merry-go-round on the left is three times that of the merry-go-round on the right.

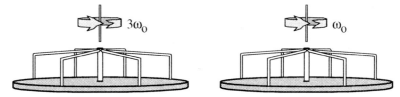

Will the net torque on the merry-go-round on the left be *greater than,* *less than,* **or** *equal to* **the one on the right?**

Explain.

NT8C-TT13: PULLEY AND BLOCK—TORQUE

A weight is tied to a rope that is wrapped around a pulley. The pulley is initially rotating counterclockwise and is pulling the weight up. The tension in the rope creates a torque on the pulley that opposes this rotation. A student makes the following comment about the torque on the pulley at the instant it stops rotating counterclockwise before starting to rotate clockwise:

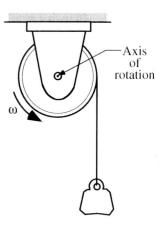

"I think the torque on the pulley at that instant will be zero since the system has stopped."

There is at least one thing wrong with this contention. Identify the error(s) and explain how to correct it (them).

NT8C-QRT14: THREE EQUAL FORCES APPLIED TO A RECTANGLE—TORQUE

Three forces of equal magnitude are applied to a 3-m by 2-m rectangle. Forces $\vec{F}_1$ and $\vec{F}_2$ act at 45° angles to the vertical as shown, while $\vec{F}_3$ acts horizontally.

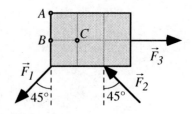

a) Is the net torque about point A clockwise, counterclockwise, or zero?
Explain how you determined your answer.

b) Is the net torque about point B clockwise, counterclockwise, or zero?
Explain how you determined your answer.

c) Is the net torque about point C clockwise, counterclockwise, or zero?
Explain how you determined your answer.

NT8C-CT15: FISHING ROD—WEIGHT OF TWO PIECES

An angler balances a fishing rod on her finger as shown.

If she were to cut the rod along the dashed line, would the weight of the piece on the left-hand side be *greater than*, *less than*, or *equal to* the weight of the piece on the right-hand side?

Explain your reasoning.

NT8C-RT16: IDENTICAL HORIZONTAL BOARDS—TORQUE ABOUT THE END

In each case below, a wooden board supports one or two boxes. The boards are uniform and have the same mass. Each board is supported by vertical posts near its ends. The masses and locations of the boxes are shown, with distances given from the post on the left end.

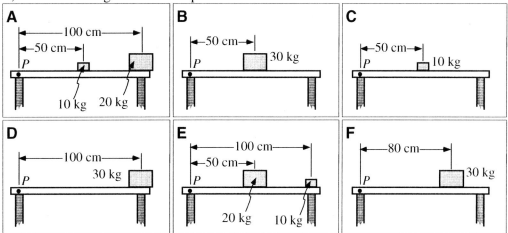

Rank these situations on the basis of the clockwise torque produced by the box(es) about point P directly above the left post.

Greatest   1 _____ 2 _____ 3 _____ 4 _____ 5 _____ 6 _____ Least

OR, The clockwise torque is the same but not zero for all these arrangements.   ___

OR, The clockwise torque is zero for all these arrangements.   ___

OR, We cannot determine the ranking for the clockwise torques in these arrangements.   ___

Please explain your reasoning.

NT8C-RT17: Horizontal Board—Clockwise Torque about Left End of the Board

In each case below, a wooden board supports a box. The boards are uniform and have the same length, but they have different masses as shown. In each case, the center of mass of the board is 50 cm from the post on the left end. Each board is supported by vertical posts near its ends. The masses and locations of the boxes are shown, with distances given from the post on the left end.

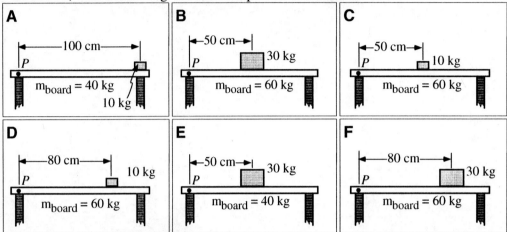

Rank these situations on the basis of the clockwise torque produced by the board and box about the left end of the board.

Greatest   1 _____ 2 _____ 3 _____ 4 _____ 5 _____ 6 _____ Least

OR, The clockwise torque is the same but not zero for all these arrangements.   ____

OR, The clockwise torque is zero for all these arrangements.   ____

OR, We cannot determine the ranking for the clockwise torques in these arrangements.   ____

Please explain your reasoning.

NT8C-RT18: Suspended Signs—Torque

The figures show six signs that are suspended from equal length rods on the side of a building. For each case, the mass of the sign compared to the mass of the rod is small and can be ignored. The mass of the sign is given in each figure. In cases B and D, the rod is horizontal; in the other cases, the angle that the rod makes with the vertical is given. The rope supporting the signs is 50 cm long in cases A, B, and C and 1 meter long in cases D, E, and F.

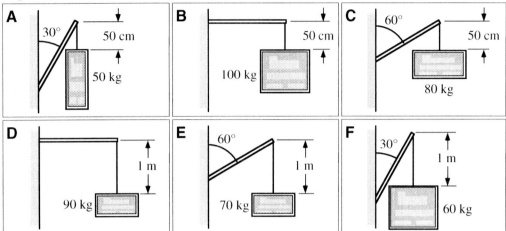

Rank these situations on the basis of the magnitude of the torque the signs exert about the point at which the rod is attached to the side of the building.

Greatest   1 _____ 2 _____ 3 _____ 4 _____ 5 _____ 6 _____ Least

OR, The torque is the same but not zero for all these arrangements.   ____

OR, The torque is zero for all these arrangements.   ____

OR, We cannot determine the ranking for the torques in these arrangements.   ____

Please explain your reasoning.

NT8C-RT19: Hexagon—Torque about Center

Four forces act on a plywood hexagon as shown in the diagram. The sides of the hexagon each have a length of 1 meter.

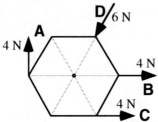

Rank the magnitude of the torque applied about the center of the hexagon by each force.

Greatest   1 _____ 2 _____ 3 _____ 4 _____   Least

OR, The magnitude of the torque due to each force is the same, but not zero.   ___

OR, The magnitude of the torque due to each force is zero.   ___

OR, We cannot determine the ranking of the magnitude of the torques.   ___

Please explain your reasoning.

NT8C-QRT20: Balance Beam—Motion After Release

Five identical keys are suspended from a balance, which is held horizontally as shown. The two keys on the left are attached to the balance 6 centimeters from the pivot and the three keys on the right are attached 5 centimeters from the pivot.

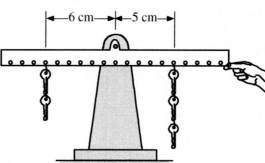

What will happen when the person lets go of the balance beam?

Explain.

NT8C-CT21: ROTATING DISCS—ANGULAR ACCELERATION, TORQUE, AND ANGLE

A net torque was applied to two discs that were initially at rest until they reached a final angular velocity. Included in the diagram below is the final angular velocity, the moment of inertia of each disc, and the time it took each disc to reach its final angular velocity. After they reach their final angular velocity, the net torque on each disc is reduced to zero, and the discs continue rotating clockwise.

Initial angular velocity: 0
Final angular velocity: ω_{Af}
Moment of inertia: $I_A = 2I_B$
Time to reach final angular velocity: Δt_A

Disc A

Initial angular velocity: 0
Final angular velocity: $\omega_{Bf} = 2\omega_{Af}$
Moment of inertia: I_B
Time to reach final angular velocity: $\Delta t_B = 2\Delta t_A$

Disc B

(1) While the discs were accelerating, was the magnitude of the angular acceleration *greater for Disc A*, *greater for Disc B*, or *the same for both discs*?

Explain.

(2) While the discs were accelerating, was the magnitude of the net torque acting on the disc *greater for Disc A*, *greater for Disc B*, or *the same for both discs*?

Explain.

(3) Was the number of revolutions turned in time Δt_1 after starting from rest *greater for Disc A*, *greater for Disc B*, or *the same for both discs*?

Explain.

(4) Was the number of revolutions turned in time Δt_2 after starting from rest *greater for Disc A*, *greater for Disc B*, or *the same for both discs*?

Explain.

NT8C-CT22: TWO MASSES ON A METER STICK—TORQUE AND DIRECTION OF ROTATION

A massless meter stick is free to rotate about a frictionless pin at the 30-cm mark. An 800-gram mass labeled m_1 is attached to the end of the meter stick at the zero cm mark and a 400-gram mass labeled m_2 is attached to the 90 cm mark.

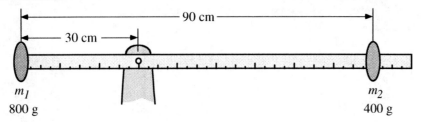

(1) If the meter stick is in the horizontal position as shown, is the magnitude of the torque about the pivot pin due to the weight of mass m_1 *greater than*, *less than*, or *equal to* the magnitude of the torque due to the weight of mass m_2?

Explain.

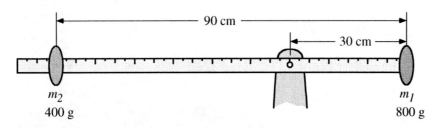

(2) If the meter stick is in the horizontal position but rotated 180° as shown above, is the magnitude of the torque about the pivot pin due to the weight of mass m_1 *greater than*, *less than*, or *equal to* the magnitude of the torque due to the weight of mass m_2?

Explain.

(3) If the meter stick is in the original horizontal position as shown at the top of the page and if it is free to rotate, will it *rotate clockwise*, *rotate counterclockwise*, or *remain stationary*?

Explain.

(4) If the meter stick is in the original horizontal position as shown at the top of the page and if it is free to rotate but the massless meter stick was replaced with a meter stick of mass 100 grams, would this meter stick *turn clockwise*, *turn counterclockwise*, or *remain stationary*?

Explain.

NT8C-CT23: TWO ROTATING GEARS—VARIOUS ASPECTS

Gear A rotates clockwise driving a second smaller gear B without slipping. (The gear teeth are not shown but have the same size and spacing.) The radius r_A for gear A is twice as large as the radius r_B for gear B.

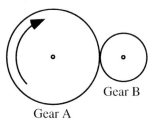

Gear B

Gear A

(1) Is the magnitude of the angular velocity of gear A *greater than, less than,* or *equal to* the magnitude of the angular velocity of gear B?

Explain.

(2) Is the magnitude of the linear velocity of a point on the edge of gear A *greater than, less than,* or *equal to* the magnitude of the linear velocity of a point on the edge of gear B?

Explain.

(3) Is the total number of gear teeth for gear A *greater than, less than,* or *the same as* the total number of gear teeth for gear B?

Explain.

(4) Is the direction of the angular velocity of gear B *the same as gear A, opposite to gear A,* or is it *impossible to compare* the direction of the angular velocities of the two gears?

Explain.

NT8C-CT24: TWO PULLEYS AND BELT—VARIOUS ASPECTS

Pulley A rotates clockwise at a constant angular velocity, driving a second larger Pulley B using a flexible belt moving around the two pulleys without slipping. The radius r_A for Pulley A is half of the radius r_B of Pulley B.

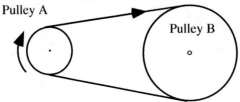

(1) **Is the magnitude of the angular velocity of Pulley A** *greater than, less than,* **or** *equal to* **the magnitude of the angular velocity of Pulley B?**

Explain.

(2) **Is the magnitude of the linear velocity of a point on the edge of Pulley A** *greater than, less than, or equal to* **the magnitude of the linear velocity of a point on the edge of Pulley B?**

Explain.

(3) **Is the direction of the angular velocity of Pulley A** *the same as, opposite to,* **or** *is it impossible to compare* **the direction of the angular velocity of Pulley B?**

Explain.

(4) **Is the direction of the angular momentum of Pulley A** *the same as* **Pulley B,** *opposite to* **Pulley B, or** *is it impossible to compare* **the direction of the angular momenta of the two pulleys?**

Explain.

NT8C-QRT25: ROTATING DISC—DIRECTION OF ANGULAR ACCELERATION, TORQUE, AND MORE

A disc with a moment of inertia of 0.2 kg·m² rotates at 300 rpm (rev/min). It took 40 s for the disc to reach this rotation rate starting from rest.

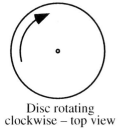

Disc rotating
clockwise – top view

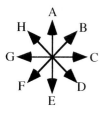

Indicate directions using the rosette above right, **J** for no direction, **K** for into page (clockwise), **L** for out of page (counterclockwise), or **M** if none of these are correct.

If the disc is rotating clockwise when viewed from above, indicate the direction of:

(1) the angular acceleration of the disc 10 seconds after starting from rest.   _____

Explain.

(2) the net torque on the disc 10 seconds after starting from rest.   _____

Explain.

(3) the angular velocity of the disc 10 seconds after starting from rest.   _____

Explain.

(4) the final angular velocity of the disc.   _____

Explain.

(5) the final angular momentum of the disc.   _____

Explain.

(6) the final kinetic energy of the disc.   _____

Explain.

nT8D-RT26: CYLINDER ON ROTATING TURNTABLE—ANGULAR MOMENTUM

Four identical metal cylinders rest on a circular horizontal turntable at various positions as shown in the top-view diagram. The turntable is rotating clockwise at a constant angular velocity.

Turntable rotating clockwise – top view

Rank the magnitude of the angular momentum of the cylinders about the axis of rotation of the turntable.

Greatest   1 _____ 2 _____ 3 _____ 4 _____   Least

OR, The magnitude of the angular momentum at all positions is the same but not zero. ____

OR, The magnitude of the angular momentum of the cylinder at all positions is zero. ____

OR, The ranking for the magnitude of the angular momentum cannot be determined. ____

Explain your reasoning.

NT8D-CT27: ROTATING DISCS—ANGULAR MOMENTUM AND ROTATIONAL KINETIC ENERGY

In case A, a disc with a moment of inertia of I_A rotates clockwise at a constant rate ω_A and in case B a disc with a moment of inertia of I_B is rotating clockwise at a constant rate ω_B as shown. I_A is two times larger than I_B and ω_B is two times larger than ω_A.

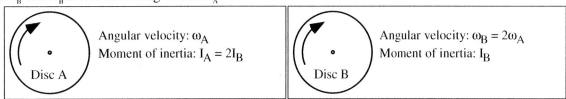

Disc A — Angular velocity: ω_A — Moment of inertia: $I_A = 2I_B$

Disc B — Angular velocity: $\omega_B = 2\omega_A$ — Moment of inertia: I_B

(1) Is the magnitude of the angular momentum of the disc in Case A *greater than, less than,* or *equal to* the magnitude of the angular momentum of the disc in Case B?

Explain.

(2) Is the kinetic energy of the disc in Case A *greater than, less than,* or *equal to* the kinetic energy of the disc in Case B?

Explain.

NT8D-CT28: MASSES ON METER STICK—ROTATIONAL INERTIA, ENERGY, & ANGULAR MOMENTUM

A massless meter stick is free to rotate about a frictionless pin at the 30-cm mark. An 800-gram mass labeled m_1 is attached to the end of the meter stick at the zero cm mark and a 400-gram mass labeled m_2 is attached to the 90 cm mark.

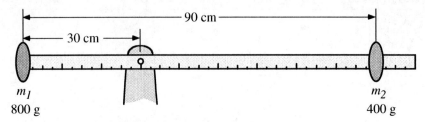

a) Is the rotational inertia of mass m_1 about the pivot pin *greater than, less than,* or *equal to* the rotational inertia of mass m_2 about the pivot point?

Explain.

b) Does the rotational inertia of the meter stick system about the pivot point *increase, decrease,* or *remain the same* if it rotates 90° from the horizontal orientation shown to a vertical orientation with mass m_2 upward?

Explain.

c) If the meter stick system is rotating at a constant angular velocity about the pivot pin, is the angular momentum of mass m_1 *greater than, less than,* or *equal to* the angular momentum of mass m_2?

Explain.

d) If the meter stick system is rotating at a constant angular velocity about the pivot pin, is the kinetic energy of mass m_1 *greater than, less than,* or *equal to* the kinetic energy of mass m_2?

Explain.

NT8D-RT29: Flat Objects with the Same Angular Velocity—Angular Momentum

Three flat objects (circular ring, circular disc, and square loop) have the same mass M and the same outer dimension (circular objects have diameters of $2R$ and the square loop has sides of $2R$). These objects are rotating with the same angular velocity. The small circle at the center of each figure represents the axis of rotation for these objects. This axis of rotation passes through the center of mass and is perpendicular to the plane of the objects.

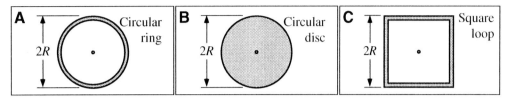

Rank the magnitude of the angular momentum of these objects about this axis of rotation.

Greatest   1 _____ 2 _____ 3 _____   Least

OR, The magnitude of the angular momentum of all of the objects is the same.   ____

OR, We cannot determine the ranking for the magnitude of the angular momentum.   ____

Please explain your reasoning.

NT8E-QRT30: SLICE OF PIZZA—ACCELERATION AND ROTATION

A slice of pizza is at rest on a frictionless table, and three people all grab for it at the same time. Each person exerts a horizontal 4-Newton force on the slice as shown to the right.

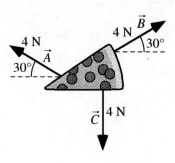

(1) Will the pizza slice accelerate?

Explain.

(2) If the pizza slice accelerates, indicate the direction of acceleration.

Explain.

(3) Will the pizza slice rotate?

Explain.

(4) If the pizza slice rotates, will it rotate clockwise or counterclockwise?

Explain.

NT8E-CT31: Pivoting Solid Disc and Ring—Torque and Angular Acceleration

A solid disc and a ring both with a radius of R, thickness t, and mass M are pivoted about a horizontal, frictionless pin through a point on their edge perpendicular to the vertical plane of the disc or ring. Both are released from rest with their centers just slightly to the right of directly above the pivot point. Consider the instants that the centers of each are in the horizontal position shown in the drawings. (The moment of inertia of the solid disc about this pivot point is $1.5MR^2$ and for the ring it is $2MR^2$.)

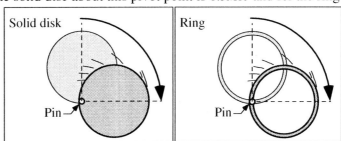

(1) At this position, will the magnitude of the torque on the solid disc about the pivot point due to its weight be *greater than, less than,* or *equal to* the magnitude of the torque on the ring about the pivot point due to its weight?

Explain.

(2) At this position, will the magnitude of the angular acceleration of the solid disc about the pivot point be *greater than, less than,* or *equal to* the magnitude of the angular acceleration of the ring due to its weight?

Explain.

nT8E-RT32: ROLLING OBJECTS RELEASED FROM REST—TIME DOWN RAMP

Four objects are placed in a row at the same height near the top of a ramp and are released from rest at the same time. The objects are (A) a 1-kg solid sphere; (B) a 1-kg hollow sphere; (C) a 2-kg solid sphere; and (D) a 1-kg thin hoop. All four objects have the same diameter, and the hoop has a width that is one-quarter its diameter. The time it takes the objects to reach the finish line near the bottom of the ramp is recorded. The moment of inertia for an axis passing through its center of mass for a solid sphere is $\frac{2}{5}MR^2$; for a hollow sphere it is $\frac{2}{3}MR^2$; and for a hoop it is MR^2.

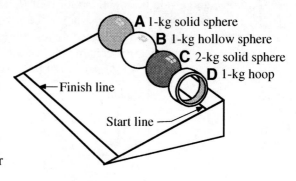

Rank the four objects from fastest (shortest time) down the ramp to slowest.

Fastest   1 _____ 2 _____ 3 _____ 4 _____   Slowest

OR, The time is the same for these objects.   ___

OR, We cannot determine the ranking for the times for these objects.   ___

Please explain your reasoning.

NT8E-CCT33: BALL ROLLING UP A HILL—FRICTION DIRECTION

A ball is rolling (without slipping) up a ramp, slowing down. Three students make the following claims about this situation:

Arlen: *"Since the ball is rolling there is no friction in this case. The only forces acting on the ball are the normal force and gravity. The reason the ball slows down is that there is a component of the weight acting down the ramp."*

Blas: *"Just because the ball is rolling doesn't mean there is no friction. There is a static force of friction that opposes the motion of the ball. The weight of the ball and the friction force both act to slow the ball down."*

Conley: *"The friction force actually points up the ramp in this case. For the ball to keep rolling without slipping, the angular speed of the ball must decrease as the speed decreases. If we think about the center of the ball as a pivot point, the only force that can produce a torque is friction, and it has to create a counterclockwise torque."*

Which, if any, of these students do you agree with?

Arlen_____ Blas _____ Conley _____ None of them_____

Explain.

NT8E-CRT34: Pulley and Weight—Angular Momentum and Torque Graphs

A weight is tied to a rope which is wrapped around a pulley. The pulley is initially rotating counterclockwise and pulling the weight upward. The weight will slow down, stop instantaneously, and then start moving downward with increasing speed.

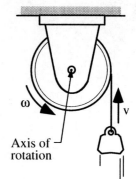

Axis of rotation

a) **On the axes below, draw a graph of the angular momentum versus time for the period from the initial instant shown above until the block comes back down to the same height.** Consider the initial angular momentum as positive.

b) **Draw a graph of the net torque on the pulley about the axis of rotation versus time for the same time period.**

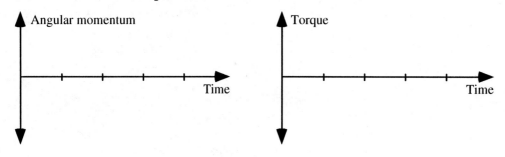

Explain.

NT8E-WWT351: Pulley with Hanging Weights—Angular Acceleration

Two pulleys with different radii (labeled *a* and *b)* are attached to one another so that they rotate together. Each pulley has a string wrapped around it with a weight hanging from it. The whole system is free to rotate about a horizontal axis in the center. The radius of the larger pulley is twice the radius of the smaller one *(b = 2a)*. A student describing this arrangement states:

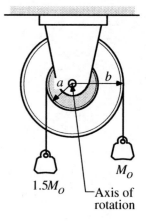

$1.5M_O$ Axis of rotation

"The larger mass is going to create a counterclockwise torque and the smaller mass will create a clockwise torque. The torque for each will be the weight times the radius, and since the radius for the larger pulley is double the radius of the smaller, and the weight of the heavier mass is less than double the weight of the smaller one, the larger pulley is going to win. The net torque will be clockwise, and so the angular acceleration will be clockwise."

What, if anything, is wrong with this contention? If something is wrong, explain how to correct it. If this contention is correct, explain why.

NT8E-CCT36: PULLEY AND MASSES—TENSION

Two masses are tied to a rope that is passed over a pulley. The pulley has mass. The mass on the right has more weight than the one on the left, and when the masses are released from rest the heavier one falls.

Four students who are asked to compare the tensions in the rope at points A and B make the following contentions:

Amir: *"I think the tension in the rope has to be the same on both sides of the pulley because there is only the one rope."*

Betsy: *"Well I disagree, because I don't see how the pulley can be turning if there isn't a torque on it, and a torque would require that there is a force acting. That force must be the difference between the two tensions. The tensions on the two sides are equal to the weights of the masses that are hanging on each side."*

Calvin: *"I think the tension in the rope is larger on the left side of the pulley because that rope has to be exerting a larger force on the left mass to pull it up. On the right side the mass exerts a larger force than the tension since that mass is moving down."*

Daria: *"Both weights are pulling at opposite ends of the rope, so the total tension in the rope is the sum of the weights. The tension is constant along any single rope as long as the mass of the rope can be ignored."*

Which, if any, of the students do you think is correct?

Amir_____ Betsy _____ Calvin _____ Daria _____ None of them_____

Explain.

NT8E-CT37: PULLEY AND HANGING MASS—ANGULAR ACCELERATION

In each case shown, a mass, initially held at rest, is hanging from a rope that is wrapped around a pulley. The pulleys have the same radius, but the pulley on the left has one-third the mass of the pulley on the right: $M_L = 1/3 M_R$. The hanging mass on the right has three times the mass as the one on the left: $3m_L = m_R$. Both masses are released and allowed to fall.

Will the angular acceleration of the pulley on the left (A) be *greater than*, *less than*, or *equal to* the pulley on the right (B)?

Explain.

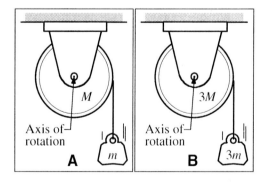

NT8E-CT38: PIVOTED ROD—INERTIA, TORQUE, ENERGY, ANGULAR ACCELERATION, & MOMENTUM

A rod with length L and mass M is pivoted about a horizontal, frictionless pin through one end. It is released from rest with its center just to the right of directly above the pivot point, so the rod rotates clockwise about the pin. Consider the time interval between the release of the rod and the instant that the rod is horizontal as shown.

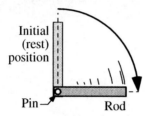

(1) Does the moment of inertia of the rod about the pivot point *increase, decrease,* **or** *remain the same* **in this time interval?**

Explain.

(2) Does the magnitude of the torque on the rod due to its weight about the pivot point *increase, decrease,* **or** *remain the same* **in this time interval?**

Explain.

(3) Does the magnitude of the angular acceleration of the rod about the pivot point *increase, decrease,* **or** *remain the same* **in this time interval?**

Explain.

(4) Does the magnitude of the angular momentum of the rod about the pivot point *increase, decrease,* **or** *remain the same* **in this time interval?**

Explain.

(5) Does the kinetic energy of the rod *increase, decrease,* **or** *remain the same* **in this time interval?**

Explain.

(6) Does the gravitational potential energy of the rod *increase, decrease,* **or** *remain the same* **in this time interval?**

Explain.

NT8F-CCT39: CHILD ON ROTATING PLATFORM—FINAL ANGULAR VELOCITY

A very low friction merry-go-round (rotating platform) with a radius of 1.5 m and moment of inertia of 50 kg·m^2 is rotating at 10 rev/min. A child with a mass of 20 kg initially sits 1 m from the center, and then moves outward toward the edge of the platform. Two students are discussing how to calculate the final angular velocity of the platform:

Arno: *"I think we have to use conservation of angular momentum. The distance from the child to the axis of rotation increases, so her moment of inertia increases. The total moment of inertia goes up, and so the angular velocity has to go down."*

Belle: *"Well, I agree that the angular velocity goes down, but I think you are going to get the wrong value. We need to use conservation of energy instead of conservation of momentum, because there is no collision here. The rotational kinetic energy is what's going to stay the same here, and as the moment of inertia increases, the angular velocity goes down."*

Which, if any, of these students do you agree with and think is correct?

Arno _____ Belle _____ Neither _____

Please explain your reasoning.

NT8F-CCT40: BULLET SHOT AT ROD—CONSERVED QUANTITIES

A bullet of mass m is shot at a hinged rod and hits the rod a distance d from the hinge. The rod was initially at rest and has a moment of inertia of I about an axis through its hinge. The bullet is fired at an angle θ to the rod, as shown in the topview, with an initial velocity of v at a distance d from the hinge. The angular speed of the rod with the bullet embedded right after the collision is ω. Several students discussing this situation state:

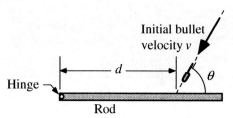

Ann: *"The angular momentum of the rod plus bullet about the hinge after the bullet is embedded is the same as the initial angular momentum of the bullet about the hinge."*

Bela: *"The kinetic energy of the bullet before the collision with the rod is equal to the rotational kinetic energy of the rod and bullet together after the bullet hits the rod."*

Carlos: *"The bullet is going straight so it does not have any angular momentum about the rod hinge before the collision, so we need to use conservation of energy. However, we can't really compare energies before and after since we do not know how much energy is lost in heat, deformation of the bullet and rod, and sound."*

Which, if any, of these students do you agree with and think is correct?

Ann _____ Bela _____ Carlos _____ None of them _____

Please explain your reasoning.

NT8F-QRT41: ENGAGING TWO CLUTCH PLATES I—ANGULAR MOMENTUM DIRECTION

Shown are two identical clutch plates rotating about the same axis with different angular velocities as shown. The upper plate will be lowered onto the lower plate ("engaging" the clutch), and after a short time they will rotate together and will have the same angular velocity.

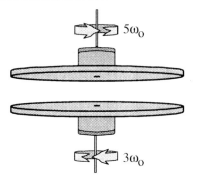

1) Before the plates touch, what is the direction of the angular momentum of the top plate?

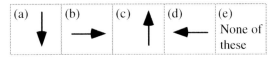

2) Before the plates touch, what is the direction of the angular momentum of the bottom plate?

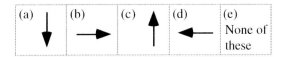

3) After the top plate is lowered onto the bottom plate, what is the direction of the angular momentum of the combined top and bottom plate system?

Explain.

4) After the top plate is lowered onto the bottom plate, what is the magnitude of the angular velocity of the combined top and bottom plate system in terms of ω_0?

Explain.

5) Does the combined system afterwards have *more*, *less*, or *the same* kinetic energy as before?

Explain.

NT8F-RT42: Bullet Shot at Rod—Angular Momentum

A bullet of mass m is shot at a hinged rod near the end of the rod. The rod was initially at rest and has a moment of inertia of I about an axis through its hinge. The bullet is fired at various angles θ to the rod, as shown in the top views below, with an initial velocity of v at a distance d from the hinge.

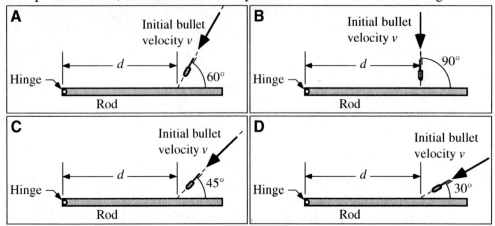

Rank the magnitude of the angular momentum of these bullets about the hinge of the rod.

Greatest   1 _____ 2 _____ 3 _____ 4 _____   Least

OR, The magnitude of the angular momentum of all the bullets will be the same but not zero. ____

OR, The magnitude of the angular momentum of all the bullets is zero. ____

OR, The magnitude of the angular momentum of the bullets is not defined. ____

OR, We cannot determine the ranking for the magnitude of the angular momentum of the bullets. ____

Please explain your reasoning.

NT8F-RT43: AIRPLANE FLYING AT CONSTANT VELOCITY AND ALTITUDE—ANGULAR MOMENTUM

An airplane is flying at a constant horizontal velocity at altitude h above an observer. The observer on the ground measures the angle between the airplane and the horizontal at four different locations as shown.

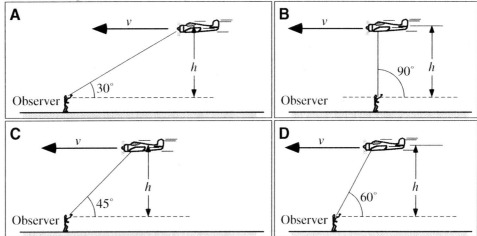

Rank the magnitude of the angular momentum of the airplane at the various locations relative to the ground observer.

Greatest 1 _____ 2 _____ 3 _____ 4 _____ Least

OR, The magnitude of the angular momentum of the airplane at the various locations will be the same but not zero. ____

OR, The magnitude of the angular momentum of the airplane at the various locations is zero. ____

OR, The magnitude of the angular momentum of the airplane is not defined. ____

OR, We cannot determine the ranking for the magnitude of the angular momentum of the airplane at the various locations. ____

Please explain your reasoning.

NT8G-RT44: Spheres Rolling Up Inclines—Height

The six figures below show solid spheres (not drawn to scale) that are about to roll up inclines without slipping. The spheres all have the same mass, but their radii as well as their linear and angular speeds at the bottom of the incline vary. Specific values are given in the figures for the linear and angular speeds at the bottom.

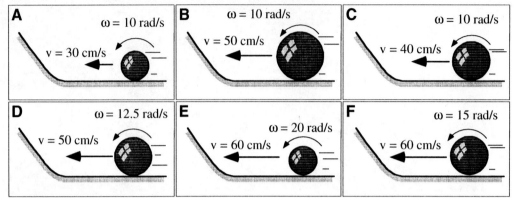

Rank these systems on the basis of the maximum height reached on the incline by each sphere.

Greatest   1 _____ 2 _____ 3 _____ 4 _____ 5 _____ 6 _____ Least

OR, The maximum heights are the same for all these objects.   ___

OR, The maximum heights are zero for all these objects.   ___

OR, We cannot determine the ranking for the maximum heights.   ___

Please explain your reasoning.

NT8G-BCT45: HOOP ROLLING UP A RAMP—BASIC BAR CHART

A thin hoop or ring with a radius of 2 m is moving so that its center of mass is initially moving at 20 m/s while also rolling without slipping at 10 rad/s along a horizontal surface. It rolls up an incline, coming to rest as shown below.

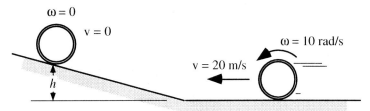

Complete the qualitative energy bar chart below for the earth-hoop system for the time between when the hoop is rolling on the horizontal surface and when it has rolled up the ramp and is momentarily at rest. Put the zero point for the gravitational potential energy at the height of the center of the hoop when it is rolling on the horizontal surface.

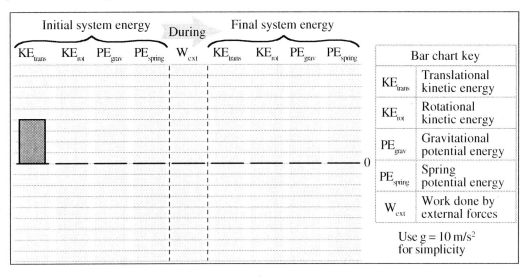

Explain.

NT8G-BCT46: Solid Disk Rolling Up a Ramp—Basic Bar Chart

A solid disk is initially rolling without slipping along a flat, level surface. It then rolls up an incline, coming momentarily to rest as shown.

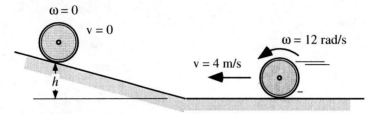

Complete the qualitative energy bar chart below for the earth-disk system for the time between when the disk is rolling on the horizontal and when it has rolled up the ramp and is momentarily at rest. Put the zero point for the gravitational potential energy at the height of the center of the hoop when it is rolling on the horizontal surface.

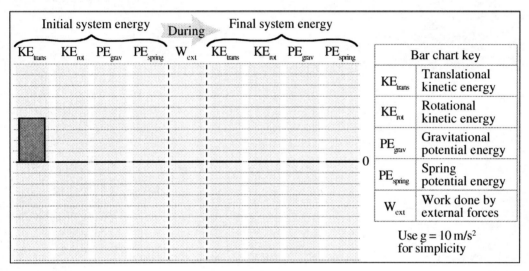

Explain.

NT8G-QRT47: Solid Sphere Rolling Along a Track—Location at Highest Point

A solid sphere rolls without slipping along a track shaped as shown at right. It starts from rest at point A and is moving vertically when it leaves the track at point B.

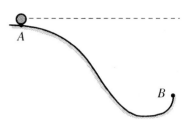

At its highest point while in the air, will the sphere be *above, below, or at the same height as* point A?

Explain your reasoning.

NT8G-WBT48: Equation I—Physical Situation

Draw and explain a physical situation where an object would be described by the following equation.

$$(4 \text{ N})(3 \text{ m}) - (2 \text{ N})(5 \text{ m}) + (2.5 \text{ N})(7 \text{ m}) = \frac{1}{3}(36 \text{ kg})(8 \text{ m})^2(0.025 \text{ rad/s}^2)$$

NT8G-WBT49: EQUATION II—PHYSICAL SITUATION

The following equation results from the application of a physical principle to an object:

$$\frac{1}{2}(4.4 \text{ kg})(1.8 \text{ m/s})^2 + \frac{1}{2}(\frac{2}{5})(4.4 \text{ kg})(0.5 \text{ m})^2(3.6 \text{ rad/s})^2 = (4.4 \text{ kg})(9.8 \text{ m/s}^2)(h)$$

Draw and explain a physical situation that would result in this equation.

NT8G-CT50: Pivoting Solid Disc and Rod—Kinetic Energy and Angular Velocity

A solid disc with a radius of R and a rod with a length of $2R$ are pivoted about a horizontal, frictionless pin through one end perpendicular to the edge of the disc or rod. They both have a mass of M and are released from rest with their centers directly above the pivot point. Consider the instant the centers of each are in the horizontal position as shown in the drawing. (The moment of inertia of the solid disc about the pivot point is $1.5MR^2$, and for the rod about its end point it is $1.33MR^2$.)

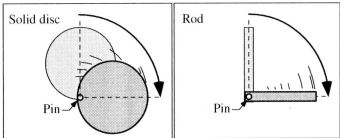

(1) Will the kinetic energy of the solid disc be *greater than, less than,* or *equal to* the kinetic energy of the rod at this position?

Explain.

(2) Will the magnitude of the angular velocity of the solid disc about the pivot point be *greater than, less than,* or *equal to* the magnitude of the angular velocity of the rod about the pivot point at this position?

Explain.

NT8G-RT51: Objects Rotating with the Same Kinetic Energy—Angular Velocity

Three flat objects (circular ring, circular disc, and square ring) have the same mass M and the same outer dimension (circular objects have diameters of $2R$ and the square ring has sides of $2R$). The small circle at the center of each figure represents the axis of rotation for these objects. This axis of rotation passes through the center of mass and is perpendicular to the plane of the objects. The objects are all rotating with the same kinetic energy about this axis.

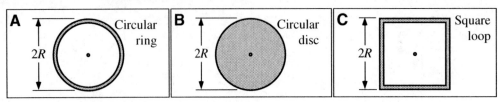

Rank the magnitude of the angular velocity of these objects about this axis of rotation.

Greatest   1 _____ 2 _____ 3 _____ 4 _____   Least

OR, The magnitude of the angular velocity of all of the objects is the same.   ___

OR, We cannot determine the ranking for the magnitude of the angular velocity.   ___

Please explain your reasoning.

NT8G-RT52: Pivoting Objects Released from Horizontal—Angular Velocity

In all four cases below, the rotating objects have a thickness t and a mass M, and are pivoted about a horizontal, frictionless pin through one end perpendicular to the plane of the object as shown. The diameter of the circular objects in cases A and B is $2R$, the length of the long rod in case C is $2R$, and the distance from the pivot to the point mass in case D is $2R$. The objects are released from rest with their centers at the same height as the pivot point. Consider the instant the centers of each are directly below the pivot as shown in the drawing. (The moments of inertia about the pivot point for the objects are $1.5MR^2$ for the disc, $2MR^2$ for the ring, $1.33MR^2$ for the rod, and $4MR^2$ for the point mass.)

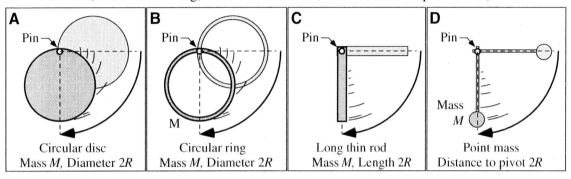

A	B	C	D
Circular disc Mass M, Diameter $2R$	Circular ring Mass M, Diameter $2R$	Long thin rod Mass M, Length $2R$	Point mass Distance to pivot $2R$

Rank the magnitude of the angular velocity of these objects about the axis of rotation when they reach the vertical position after they have rotated 90°.

Greatest 1 _____ 2 _____ 3 _____ 4 _____ Least

OR, The magnitude of the angular velocity of the objects in the vertical position is the same. ___

OR, We cannot determine the ranking for the magnitude of the angular velocity of the objects. ___

Please explain your reasoning.

NT8G-RT53: Objects Moving Down a Ramp—Maximum Height of Airborne Objects

In each case below, an object is released from rest at point *S* on a ramp at a height of 2 meters from the bottom. The objects are moving exactly vertically when they are launched from the ramp at point *T*. All of the ramps are the same shape. All of the spheres roll without slipping, and the blocks slide without friction. The shapes and masses of the objects are given in the drawing for each case.

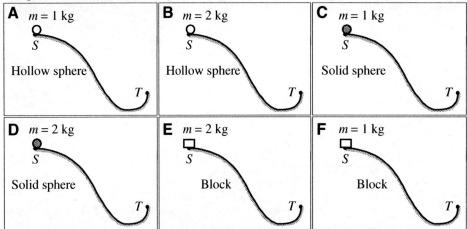

Rank the maximum height of the objects after they become airborne at point *T*.

Greatest   1 _____ 2 _____ 3 _____ 4 _____ 5 _____ 6 _____ Least

OR, The maximum height of all of the objects is the same.   ____

OR, We cannot determine the ranking for the maximum height of the objects.   ____

Please explain your reasoning.

NT8G-RT54: Moving Down a Ramp—Maximum Height on Other Side of Ramp

In each case below, a 1-kg object is released from rest on a ramp at a height of 2 meters from the bottom. All of the spheres roll without slipping, and the blocks slide without friction.

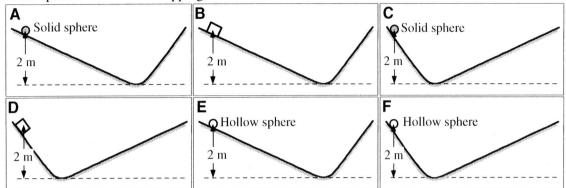

Rank the maximum height of the objects on the other side of the ramp.

Greatest   1 _____ 2 _____ 3 _____ 4 _____ 5 _____ 6 _____ Least

OR, The maximum height of all of the objects is the same.   ____

OR, We cannot determine the ranking for the maximum height of the objects.   ____

Please explain your reasoning.

NT8G-RT55: OBJECTS MOVING DOWN RAMPS—SPEED AT BOTTOM

In each case below, a 1-kg object is released from rest on a ramp at a height of 2 meters from the bottom. All of the spheres roll without slipping, and the blocks slide without friction. The ramps are identical in cases A, D, and F. The ramps in cases B, C, and E are identical and are not as steep as the others.

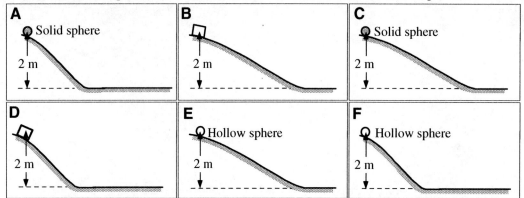

Rank these cases on the basis of the speed of the objects when they reach the horizontal surface at the bottom of the ramp.

Greatest   1 _____ 2 _____ 3 _____ 4 _____ 5 _____ 6 _____ Least

OR, The maximum speed is the same for all cases.   ____

OR, We cannot determine the ranking for the maximum speed of these objects.   ____

Please explain your reasoning.

NT8H-RT56: Stationary Horizontal Boards—Force by Left Post on Board

In each case below, uniform boards with different masses are supported by two vertical posts. Each board holds either a 10-kg box or a 30-kg box. The boards all have the same lengths. The masses of the boards and boxes as well as the distances from the left post to the center of mass of each box are given.

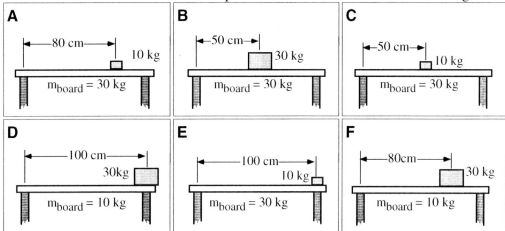

Rank these situations on the basis of the force that the left post exerts on the board.

Greatest   1 _____ 2 _____ 3 _____ 4 _____ 5 _____ 6 _____ Least

OR, The force is the same but not zero for all these arrangements.   ____

OR, The force is zero for all these arrangements.   ____

OR, We cannot determine the ranking for the forces in these arrangements.   ____

Please explain your reasoning.

NT8H-RT57: Tilted Pivoted Rods with Various Loads—Force to Hold Rods

Six identical massless rods are all supported by a fulcrum and are tilted at the same angle to the horizontal. A mass is suspended from the left end of the rod, and the rods are held motionless by a downward force on the right end. The locations of the rod with respect to the fulcrum and the values of the suspended masses are shown for each case. Each rod is marked at 1-meter intervals.

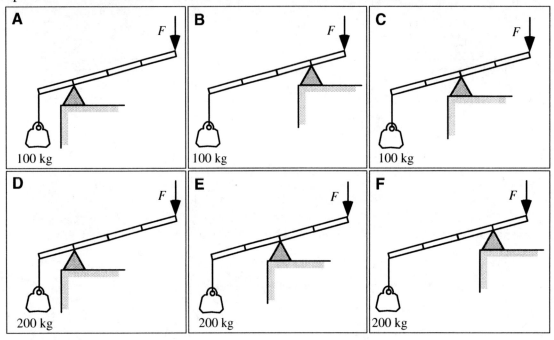

Rank the magnitude of the vertical force *F* applied to the end of the rod.

Greatest   1 _____ 2 _____ 3 _____ 4 _____ 5 _____ 6 _____ Least

OR, The vertical force applied to the end of the rod for all of the systems is the same.   ____

OR, We cannot determine the ranking for the vertical force for these systems.   ____

Please explain your reasoning.

NT8H-CT58: TILTED PIVOTED RODS WITH VARIOUS LOADS—FORCE TO HOLD RODS

In each case below, a massless rod is supported by a fulcrum. A 200-kg hanging mass is suspended from the left end of the rod by a cable. A downward force F keeps the rod at rest. The rod in Case A is 50 cm long, and the rod in Case B is 40 cm long. (Each rod is marked at 10-centimeter intervals.)

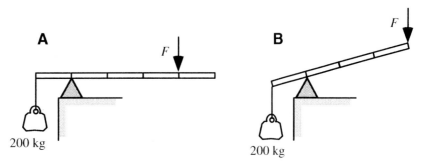

Will the magnitude of the vertical force, F, exerted on the rod in Case A be *greater than*, *less than*, or *equal to* the magnitude of the vertical force, F, exerted on the rod in Case B?

Explain.

NT8H-WBT60: TWO EQUATIONS—PHYSICAL SITUATION

The following equations result from the application of physical principles to a system:

$$T_{\text{left}} + T_{\text{right}} - (25 \text{ kg})(9.8 \text{ m/s}^2) - (60 \text{ kg})(9.8 \text{ m/s}^2) - (40 \text{ kg})(9.8 \text{ m/s}^2) = 0$$

$$(25 \text{ kg})(9.8 \text{ m/s}^2)(4 \text{ m}) + (60 \text{ kg})(9.8 \text{ m/s}^2)(6 \text{ m}) + (40 \text{ kg})(9.8 \text{ m/s}^2)(9 \text{ m}) - (T_{\text{right}})(12 \text{ m}) = 0$$

Draw a physical situation that would result in these equations and explain how your drawing is consistent with the equations.

NT8H-RT61: Horizontal Pivoted Rods with Loads at Various Locations—Force to Hold

Six identical 2-meter massless rods are supporting identical 12-Newton weights. In each case, a vertical force F is holding the rods and the weights at rest. The left end of each rod is held in place by a frictionless pin. The rods are marked at half-meter intervals.

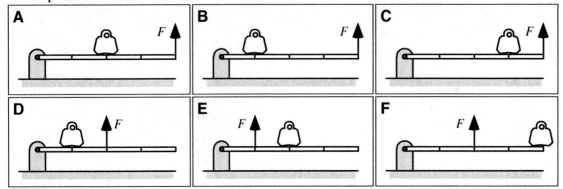

Rank the magnitude of the vertical force F applied to the rods.

Greatest   1 _____ 2 _____ 3 _____ 4 _____ 5 _____ 6 _____ Least

OR, The applied vertical force for all of the systems will be the same.   ____

OR, We cannot determine the ranking for the applied vertical force for all of the systems.   ____

Please explain your reasoning.

NT8H-LMCT62: HORIZONTAL PIVOTED BOARD WITH LOAD I—FORCE TO HOLD BOARD

A 100-N weight is placed on a massless board a distance L_1 from a frictionless pin. A force F is applied vertically to the end of the board, a distance L_2 from the pivot.

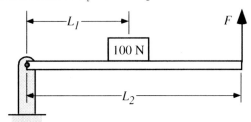

Identify from choices (a)-(e) how each change described below will affect the magnitude of the support force (F) at the end of the board needed to keep the system in equilibrium.

Compared to the case above, this change will:

(a) *increase* the magnitude of the support force *(F)* at the end of the board.

(b) *decrease* the magnitude of the support force *(F)* at the end of the board but not to zero.

(c) *decrease* the magnitude of the support force *(F)* at the end of the board to zero.

(d) *have no effect* on the magnitude of the support force *(F)* at the end of the board.

(e) *have an indeterminate* effect on the magnitude of the support force *(F)* at the end of the board.

Each of these modifications is the only change to the initial situation shown in the diagram above.

1) The 100-N weight is moved to a position closer to the support force (F) at the right end of the board. _____

Explain.

2) The 100-N weight is moved to a position closer to the frictionless pin at the left end of the board. _____

Explain.

3) The weight is increased to 150 N. _____

Explain.

4) With the distance L_1 unchanged, the length L_2 is increased (the board is made longer) and the support force (F) is still at the end of the board. _____

Explain.

5) The 100-N weight is moved to a position closer to the frictionless pin at the left end of the board, and the board is made longer, but the distance L_2 to the support force is unchanged. _____

Explain.

6) The support force (F) is moved to a position closer to (but still to the right of) the 100-N weight. _____

Explain.

NT8H-LMCT63: HORIZONTAL PIVOTED BOARD WITH LOAD II—FORCE TO HOLD BOARD

A 100-N weight is placed on a massless board a distance L_1 to the left of frictionless pin. A vertical downward force F is applied to the other side of the board a distance of L_2 from the pin as shown. The system is at rest.

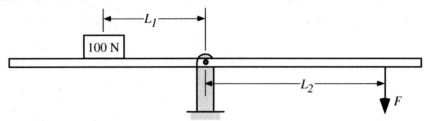

Identify from choices (a)-(e) how each change described below will affect the magnitude of the applied force (F) on the right side of the board needed to keep the system in equilibrium.

Compared to the case above, this change will:

 (a) *increase* the magnitude of the support force *(F)* on the board.

 (b) *decrease* the magnitude of the support force *(F)* on the board but not to zero.

 (c) *decrease* the magnitude of the support force *(F)* on the board to zero.

 (d) *have no effect* on the magnitude of the support force *(F)* on the board.

 (e) *have an indeterminate* effect on the magnitude of the support force *(F)* on the board.

Each of these modifications is the only change to the initial situation shown in the diagram above.

1) The 100-N weight is moved to a position closer to the pin.   _____
Explain.

2) The support force (F) is moved to a position closer to the pin.   _____
Explain.

3) The weight is decreased to 50 N.   _____
Explain.

4) The support force (F) is moved to the right end of the board.   _____
Explain.

5) The board is made longer but the support force (F) remains at the same location.   _____
Explain.

6) The 100-N weight and the support force (F) are both moved to positions closer to the pin.   _____
Explain.

NT8H-CT64: Mass Suspended from a Horizontal Beam—Tension and Torque

A beam is supported by a frictionless pin so that it is horizontal. A mass is suspended from the right end of the beam, and the left end is tethered by an anchored cable. Both cases below are identical, except for the location of the anchor point of the cable.

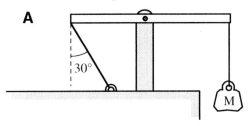 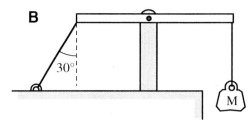

(1) Is the magnitude of the torque due to the cable about the pivot pin *greater in case A, greater in case B,* or *the same in both cases?*

Explain.

(2) Is the tension in the cable *greater in case A, greater in case B,* or *the same in both cases?*

Explain.

NT8H-CT65: Mass Suspended from a Tilted Beam—Tension and Torque

A beam is supported by a frictionless pin so that it makes a 20° angle with the horizontal. A mass is suspended from the right end of the beam, and the left end is tethered by an anchored cable. Both cases below are identical, except for the location of the anchor point of the cable.

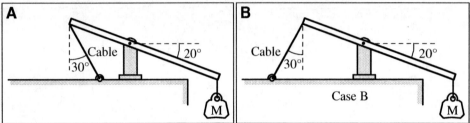

(1) Is the magnitude of the torque due to the cable about the pivot point *greater in case A*, *greater in case B*, or *the same in both cases*?

Explain.

(2) Is the tension in the cable *greater in case A*, *greater in case B*, or *the same in both cases*?

Explain.

NT8H-CCT66: Hanging Traffic Light—Tension

An electrician is suspending a traffic light between two poles using cables as shown. Three physics students who are watching make the following contentions:

Anders: *"For this arrangement, the tension has to be the same in both cables, since they each support half the weight of the traffic light."*

Baji: *"I think the tension in the left cable has to be larger because it is more vertical. The vertical component of its tension has to be larger than for the right-hand cable."*

Caitlin: *"I think the tension in the left cable actually has to be less than the tension in the right cable. It is pulling more vertically, so it has an easier job of lifting its half of the traffic light. The cable on the right still has to lift its half, but since it is mostly pulling horizontally it has to have a really large tension to get a large enough vertical component."*

Which, if any, of the students do you think is correct?

Anders_____ Baji _____ Caitlin _____ None of them_____

Explain.

NT8H-RT67: Horizontal Uniform Rods—Force of Pivot on Rods

In each case below, a uniform rod is attached to a support at one end with a frictionless pivot. At the other end of each rod, an upward support force F is applied that keeps the rod horizontal. The lengths and masses of the rods are given. The rods are marked in equal intervals of 1 meter.

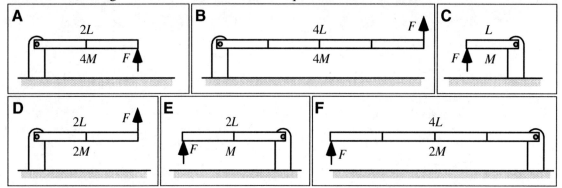

Rank the magnitude of the upward force that the pivot pin exerts on the rod.

Greatest 1_____ 2_____ 3_____ 4_____ 5_____ 6_____ Least

OR, The pivot pin exerts the same upward force in all cases.   ____

OR, We cannot determine the ranking for force that the pivot pin exerts.   ____

Please explain your reasoning.

NT8H-CT68: HANGING UP TWO CARPENTER'S SQUARES—FORCE BY SUPPORT DOWELS

Two identical carpenter's squares are supported on a wall by dowels as shown.

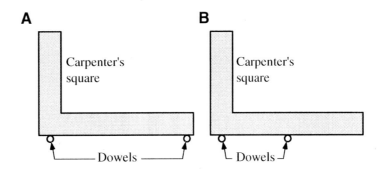

(1) Is the force on the carpenter's square by the right dowel *greater in case A, greater in case B,* **or** *the same in both cases?*

Explain.

(2) Is the force on the carpenter's square by the left dowel *greater in case A, greater in case B,* **or** *the same in both cases?*

Explain.

NT9A-RT1: MASS ON A VERTICAL SPRING—TIME

The figures below show six identical masses attached to springs and hung vertically. The masses are pulled down various distances and then released. The spring constant (k), which measures the stiffness of the spring, and the distance (d) that the mass is pulled down are given for each case in the figures.

A	B	C	D	E	F
d = 60 cm	d = 90 cm	d = 50 cm	d = 90 cm	d = 50 cm	d = 50 cm
k = 240 N/m	k = 180 N/m	k = 150 N/m	k = 150 N/m	k = 240 N/m	k = 180 N/m

Rank these situations on the basis of the time for the mass to get from the maximum distance below the equilibrium point to the maximum distance above the equilibrium point.

Greatest   1 _____ 2 _____ 3 _____ 4 _____ 5 _____ 6 _____   Least

OR, The times are the same for all these cases.   ___

OR, The times are zero for all these cases.   ___

OR, We cannot determine the ranking for the times of these cases. ___

Please explain your reasoning.

The figures below show six metal pieces attached to identical springs and hung vertically. The metal pieces are pulled down various distances and released. In each figure the kinetic energy of the metal piece as it passes through the equilibrium point and the mass of the piece are given.

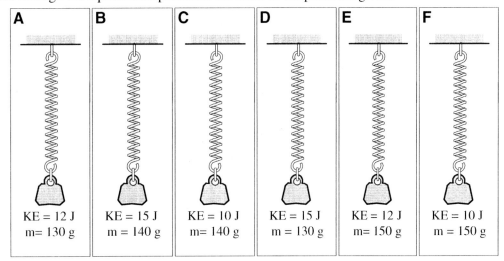

A	B	C	D	E	F
KE = 12 J	KE = 15 J	KE = 10 J	KE = 15 J	KE = 12 J	KE = 10 J
m= 130 g	m = 140 g	m= 140 g	m = 130 g	m= 150 g	m = 150 g

Rank these situations on the basis of the amplitude of the oscillation, i.e., the maximum distance the metal pieces move down from the equilibrium point.

Greatest   1 _____ 2 _____ 3 _____ 4 _____ 5 _____ 6 _____ Least

OR, The amplitudes are the same for all these situations.   ____

OR, The amplitudes are zero for all these situations.   ____

OR, We cannot determine the ranking for these situations.   ____

Please explain your reasoning.

NT9A-RT3: MASS ON HORIZONTAL SPRING SYSTEMS—OSCILLATION FREQUENCY

The figures below show systems containing a block resting on a frictionless surface and attached to the end of a spring. The springs are stretched to the right by a distance given in each figure and then released from rest. The blocks oscillate back and forth. The mass and force constant are given for each system.

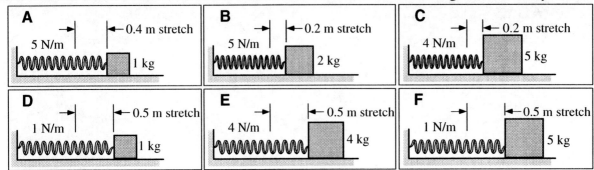

Rank the systems on the basis of the frequency of the vibratory motion.

Greatest   1 _____ 2 _____ 3 _____ 4 _____ 5 _____ 6 _____ Least

OR, All of the frequencies will be the same for each system.   ___

OR, We cannot determine the ranking for the frequencies of these oscillations.   ___

Please explain your reasoning.

NT9A-RT4: Sphere on a String—String Length

The six figures below show metal spheres hung on the ends of strings. The spheres have been pulled to the side and released so that they are swinging back and forth. For each sphere-string system the diagrams give the mass of the sphere, the frequency of the swing, and how far, in terms of the angle from the vertical, that the spheres were initially pulled to the side.

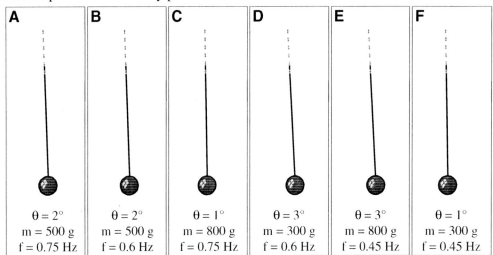

A	B	C	D	E	F
$\theta = 2°$	$\theta = 2°$	$\theta = 1°$	$\theta = 3°$	$\theta = 3°$	$\theta = 1°$
m = 500 g	m = 500 g	m = 800 g	m = 300 g	m = 800 g	m = 300 g
f = 0.75 Hz	f = 0.6 Hz	f = 0.75 Hz	f = 0.6 Hz	f = 0.45 Hz	f = 0.45 Hz

Rank these systems on the basis of the length of the string.

Greatest   1 _____ 2 _____ 3 _____ 4 _____ 5 _____ 6 _____ Least

OR, The string lengths are the same for all these systems.   ____

OR, We cannot determine the ranking for the lengths of the strings.   ____

Please explain your reasoning.

NT9A-CT5: MASS HANGING ON A SPRING—PERIOD OF OSCILLATION

A mass *M* hangs from a spring as shown. If the mass is pulled down a small distance *y* from its equilibrium position and released, the system will oscillate up and down.

a) If the mass is instead pulled down a distance *2y*, will the period of oscillation for the spring-mass system be *greater than, less than,* or *equal to* the period of oscillation of the same spring-mass system when it was pulled down a distance *y*?

Explain.

b) If this spring-mass system were given a larger total energy, would the period of oscillation (the time for one complete cycle of motion) *increase, decrease,* or *remain the same?*

Explain.

c) If an additional mass *m* is added to the original mass *M*, will the period of oscillation for the spring-mass system be *greater than, less than,* or *equal to* the period of the spring-mass system with only mass *M*?

Explain.

NT9A-RT6: Cart and Springs—Oscillation Frequency

The springs shown below have the same unstretched length and the same spring constant k. All of the carts are identical and are on horizontal frictionless surfaces. In case C, there are two springs attached end-to-end. The carts are given a small horizontal displacement and then released.

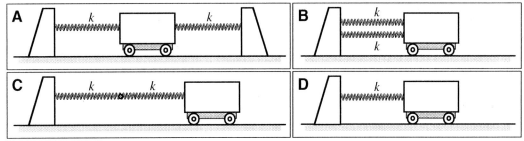

Rank these cases on the basis of the oscillation frequency of the carts.

Greatest   1 _____ 2 _____ 3 _____ 4 _____   Least

OR, The frequencies are the same for all these cases. ____

OR, We cannot determine the ranking for the frequencies of these systems. ____

Please explain your reasoning.

NT9A-RT7: Mass and Spring Systems—Oscillation Frequency

The mass-spring system shown in figure A consists of a spring with a spring constant k and unstretched length L connected to a mass M. The cart is resting on a horizontal frictionless surface. If the mass is pulled to one side a small distance and released, the mass will oscillate back and forth with an amplitude A and frequency f_o. Cases B, C, and D shown below are variations of this system.

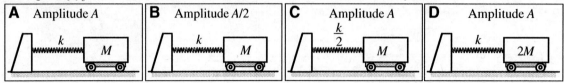

Rank these cases shown above on the basis of their oscillation frequency.

Greatest 1 _____ 2 _____ 3 _____ 4 _____ Least

OR, The frequencies are the same for all four cases. ___

OR, We cannot determine the ranking for the frequencies. ___

Please explain your reasoning.

NT9A-RT8: Simple Pendula—Oscillation Frequency

A simple pendulum consists of a mass M attached to a massless string of length L as shown in figure A below. If the mass is pulled to one side a small distance and released, it will swing back and forth with a frequency f_o and swing amplitude A. Cases B, C, and D shown below are variations of this system.

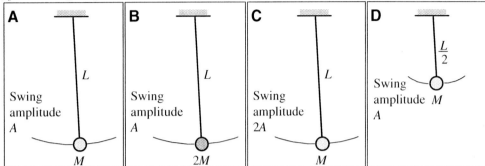

Rank these cases on the basis of the oscillation frequency of the masses.

Greatest 1 _____ 2 _____ 3 _____ 4 _____ Least

OR, The frequencies are the same for all these cases. ____

OR, We cannot determine the ranking for the frequencies. ____

Please explain your reasoning.

NT9A-QRT9: POSITION VS. TIME GRAPH OF A CART ATTACHED TO A SPRING—MASS & PERIOD

A frictionless cart of mass m is attached to a spring with spring constant k. When the cart is displaced from its rest position and released, it oscillates with a period τ that is given by

$$\tau = 2\pi\sqrt{m/k}.$$

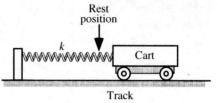

The graph of the position of the cart as a function of time is shown below for Experiment A. Graphs for two other experiments are shown below this. The same spring is used in all three experiments.

(1) Compared to Experiment A, in Experiment B the cart has

a) *twice* as much mass.

b) *four times* as much mass.

c) *one-half* the mass.

d) *one-fourth* the mass.

e) *the same* mass.

Explain.

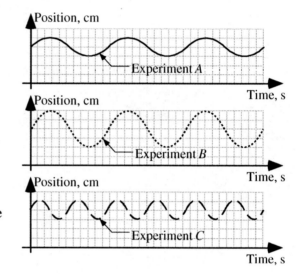

(2) Compared to Experiment A, in Experiment C the cart has

a) *twice* as much mass.

b) *four times* as much mass.

c) *one-half* the mass.

d) *one-fourth* the mass.

e) *the same* mass.

Explain.

(3) Suppose that for a fourth experiment (Experiment D), the mass used in Experiment A was doubled and the spring was replaced with a spring with double the spring constant. The period in Experiment D would be

a) *the same* as the period in Experiment A.

b) *double* the period in Experiment A.

c) *four times* the period in Experiment A.

d) *one-half* the period in Experiment A.

e) *one-fourth* the period in Experiment A.

Explain.

NT9A-LMCT10: Swinging Heavy Rod—Frequency

A uniform rod with a mass M and a length L is suspended vertically by a frictionless pin at one end as shown. If the rod is pulled to one side a small distance and released, it will swing back and forth with a frequency f_o.

Identify from choices (a)-(d) how each change described below will affect the frequency of the swinging rod.

Compared to the case above, this change will:

   (a) *increase* the frequency of the swinging rod.

   (b) *decrease* the frequency of the swinging rod.

   (c) *have no effect* on the frequency of the swinging rod.

   (d) *have an indeterminate* effect on the frequency of the swinging rod.

Each of these modifications is the only change to the initial situation described above.

(1) The mass of the rod is increased.   _____

Explain.

(2) The length of the rod is increased.   _____

Explain.

(3) The rod is pulled to one side a little more and released.   _____

Explain.

(4) The length of the rod is doubled to 2*L* and the mass is decreased to *M*/2.   _____

Explain.

(5) The length of the rod is halved to *L*/2 and the mass is increased to 2*M*.   _____

Explain.

(6) The rod is swinging on the moon (smaller gravitational forces).   _____

Explain.

A cart attached to a spring is displaced from equilibrium and then released. A graph of displacement as a function of time for the cart is shown below. There is no friction.

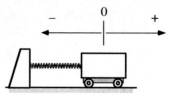

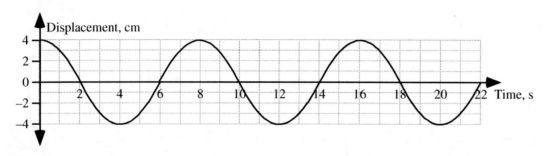

(a) What is the mathematical expression for the displacement of the cart as a function of time?

Explain.

(b) What is the mathematical expression for the velocity of the cart as a function of time?

Explain.

(c) What is the mathematical expression for the acceleration of the mass as a function of time?

Explain.

NT9A-CRT12: VELOCITY VS. TIME GRAPH—FREQUENCY AND PERIOD

A cart attached to a spring is displaced from equilibrium and then released. A graph of velocity as a function of time for the cart is shown below. There is no friction.

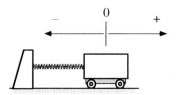

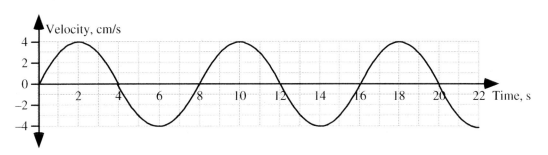

(a) What is the period of the motion for this cart?

Explain.

(b) What is the frequency of the motion for this cart?

Explain.

(c) In which direction was the cart displaced from equilibrium before it was released?

Explain.

NT9A-CRT13: VELOCITY VS. TIME—VELOCITY AND ACCELERATION EXPRESSION

A mass attached to a spring is displaced from equilibrium and then released. A graph of velocity of the mass as a function of time is shown below. There is no friction in the system.

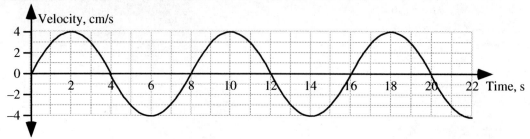

(a) What is the mathematical expression for the velocity of the mass as a function of time?

Explain.

(b) What is the mathematical expression for the acceleration of the mass as a function of time?

Explain.

(c) What is the mathematical expression for the displacement of the mass as a function of time?

Explain.

NT9A-CT14: Mass Hanging on a Spring—Period of Oscillation

A spring has a length L when a hanging mass M is attached. If the mass M is pulled down a small distance from its equilibrium position and released, the system will oscillate up and down with an initial amplitude A, frequency f, and period T (the time for one complete cycle of motion). If a mass m is added to the hanging mass, it stretches the spring a distance l longer.

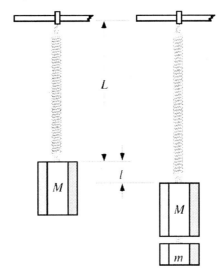

(1) **Will the period of oscillation for the spring-mass system with mass M be *greater than*, *less than*, or *equal to* the period of the spring-mass system with mass $(m + M)$? Assume that they both are released so that they oscillate with the same amplitude.**

Explain.

(2) **If M is pulled down a distance of $2l$ and released, will the period of oscillation for this spring-mass system be *greater than*, *less than*, or *equal to* the period of the same spring-mass system that is pulled down a larger distance $3l$ and released?**

Explain.

NT9A-LMCT15: Mass Connected to a Spring—Frequency

A mass-spring system consists of a spring with a spring constant (or stiffness) k and unstretched length L, connected to a cart of mass M resting on a horizontal frictionless surface as shown. If the cart is pulled to one side a small distance and released, it will oscillate back and forth with amplitude A and frequency f.

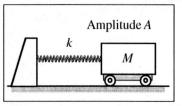

Identify from choices (a) – (d) how each change described below will affect the frequency of the oscillating mass-spring system.

Compared to the case above, this change will:

    (a) *increase* the frequency of the system.

    (b) *decrease* the frequency of the system.

    (c) *have no effect* on the frequency of the system.

    (d) *have an indeterminate* effect on the frequency of the system.

Each of these modifications is the only change to the initial situation described above.

(1) The mass is increased. _____

Explain.

(2) The spring constant or stiffness is increased. _____

Explain.

(3) The mass is pulled a little farther and then released. _____

Explain.

(4) The spring constant is doubled to 2k and the mass is reduced to M/2. _____

Explain.

(5) The amplitude is increased and the mass is increased. _____

Explain.

(6) The mass-spring system is oscillating on the moon (smaller gravitational forces). _____

Explain.

NT9B-LMCT16: Oscillation Graph of Displacement vs. Time—Kinematic Quantities

A cart attached to a spring is displaced from equilibrium and then released. A graph of displacement as a function of time for the cart is shown below. There is no friction. Eight points are labeled A – H in the graph.

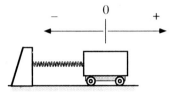

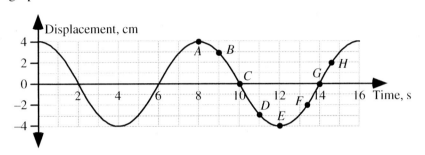

For each question below, choose from the labeled points above or state "none."

1. At which point or points is the acceleration positive?
Explain.

2. At which point or points does the cart have zero velocity but nonzero net force?
Explain.

3. At which point or points is the net force on the cart equal to zero?
Explain.

4. At which point or points are the acceleration, velocity, and displacement all positive?
Explain.

5. At which point or points is the acceleration nonzero and opposite in sign to the position?
Explain.

6. At which point or points is the velocity nonzero and opposite in sign to the acceleration?
Explain.

NT9B-CCT17: Oscillating Cart—Period

A frictionless cart of mass M is attached to a spring with spring constant k. When the cart is displaced 6 centimeters from its rest position and released, it oscillates with a period of 2 seconds.

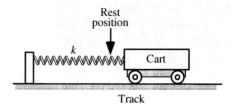

Four students are discussing what would happen to the period of oscillation if the original cart with mass M was displaced 12 cm from its rest position instead of 6 cm and again released:

Ava: *"Since the spring is stretched more, the force will be greater, causing a greater acceleration and greater speeds overall. Since the cart is moving faster, the time will go down, probably to 1 second since the force is doubled."*

Barb: *"The cart has farther to go now, and so it's going to take longer to make a complete cycle. It's going to go farther on both sides of the rest position, so the round-trip is 48 cm instead of 24 cm. The period is going to double."*

Charles: *"The cart has four times as much energy, and the conjugate variable for energy is time according to the Heisenberg uncertainty principle. The energy quadruples when the spring stretch is doubled, and so the time must be only one-quarter as much. The period will be one-half second."*

Dorothy: *"Stretching the spring twice as far means that k is twice as big. And the period is 2 pi times the square root of the spring constant divided by the mass. Doubling the spring constant and leaving the mass alone is going to double what's inside the square root, and after we take the square root we get a period of 2 seconds times the square root of 2, or 1.414 seconds."*

Which, if any, of these students do you agree with?

Ava _____ Barb _____ Charles _____ Dorothy _____ None of them _____

Please explain your reasoning.

NT9B-QRT18: OSCILLATING MASS-SPRING GRAPH OF DISPLACEMENT VS. TIME—DIRECTIONS

A cart attached to a spring is displaced from equilibrium and then released. A graph of displacement as a function of time for the cart is shown. There is no friction. Points are labeled *A* – *H* in the graph.

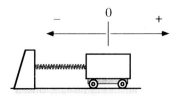

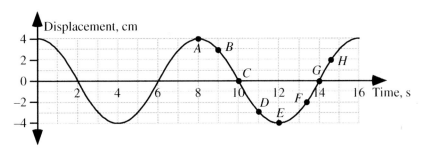

For each labeled point above, identify if the vector quantity listed below is in the positive (+) direction, negative (–) direction, or is zero (0) for no direction.

Point	Acceleration	Velocity	Displacement	Net Force
A				
B				
C				
D				
E				
F				
G				
H				

NT9C-RT19: Oscillating Chip Testing System—Maximum Force

A small electronic chip is being tested to withstand forces up to 10 times its weight. To test the chip, it is fastened to a cart. The cart is attached to a spring as shown, and moves from side to side with simple harmonic motion at a frequency f and amplitude A. The system is frictionless, and the masses given are of the cart with the chip attached.

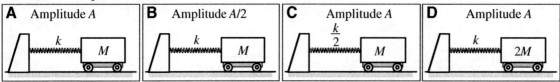

A Amplitude A	**B** Amplitude $A/2$	**C** Amplitude A $\frac{k}{2}$	**D** Amplitude A
k M	k M	M	k $2M$

Rank the cases shown above on the basis of the maximum force that the spring exerts on the cart.

Greatest   1 _____ 2 _____ 3 _____ 4 _____   Least

OR, All of the maximum force will be the same for each system.   ____

OR, We cannot determine the ranking for the maximum force for these systems.   ____

Please explain your reasoning.

NT9C-CCT20: Mass on a Spring—Acceleration

A mass is oscillating up and down at the end of a spring. Three students state:

Aileen: *"I think the acceleration of the mass will be largest when it is at the end of its oscillations turning around. That's where the spring is stretched the most."*

Ben: *"No, I don't see how that can be since its velocity is zero at that point, so its acceleration has to be zero also."*

Carl: *"I disagree. The acceleration is largest when the mass is halfway between the middle and the end because that is where its speed is changing the most."*

Which, if any, of these students do you think is correct?

Aileen _____ Ben _____ Carl _____ None of them_____
Please explain your reasoning.

NT9C-WBT21: Acceleration of an Object—Physical Situation

Describe a physical situation where an object would have the following acceleration.

$$a(t) = (0.4 \, \frac{\text{m}}{\text{s}^2}) \sin(2t)$$

NT9D-LMCT22: DISPLACEMENT VS. TIME—ENERGY QUANTITIES

A cart attached to a spring is displaced from equilibrium and then released. A graph of displacement as a function of time for the cart is shown. There is no friction. Points are labeled *A – H* in the graph.

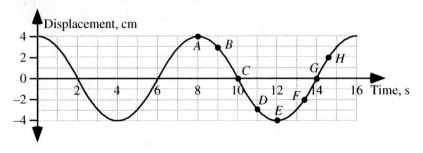

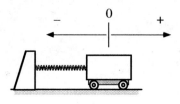

For each question below, choose from the labeled points above or state "none" for the mass-spring-earth system.

1. At which point or points are the spring potential energy and the cart's kinetic energy both at their maximum values?
Explain.

2. At which point or points is the kinetic energy equal to zero?
Explain.

3. At which point or points is the total energy at its maximum value?
Explain.

4. At which point or points is the spring potential energy negative?
Explain.

5. At which point or points is the kinetic energy positive?
Explain.

6. At which point or points is the kinetic energy at its maximum value and the spring potential energy at its minimum value?
Explain.

7. At which point or points is the kinetic energy at its minimum value and the spring potential energy at its maximum value?
Explain.

NT9D-CCT23: Mass Oscillating on a Vertical Spring—Energy

A mass hanging on a vertical spring is pulled down a distance d and released. The mass undergoes simple harmonic motion. Three physics students make the following contentions about this situation:

Alexandra: *"The maximum kinetic energy of this mass-spring system is fixed by the properties of the system and does not depend on how far down the mass is pulled. How far the mass is pulled will only affect the frequency of the oscillations."*

Bao: *"No, that can't be right since increasing the amplitude, or how far down it is pulled, increases the potential energy of the system. And I don't think the amplitude has any effect on the frequency."*

Chung: *"I agree in part with both of you. I think the amplitude does affect the maximum kinetic energy, but I also think it affects the frequency of the oscillations."*

Which, if any, of these three students do you agree with and think is correct?

Alexandra _____ Bao _____ Chung _____ None of them _____

Please explain your reasoning.

NT9D-LMCT24: OSCILLATION GRAPH FOR A HANGING MASS ON A SPRING—ENERGY QUANTITIES

A mass attached to a spring is suspended as shown. When the mass is at rest the spring is 12 centimeters long. The unstretched length of the spring is 8 centimeters. The mass is pulled down so that the spring length is 16 centimeters and then released. A graph of the spring length as a function of time is shown. Eight points are labeled *A – H* in the graph.

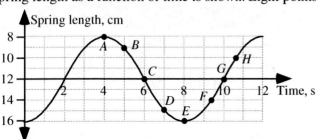 

For each question below, choose from the labeled points above or state "none."

1. At which point or points is the spring potential energy at its maximum value?

Explain.

2. At which point or points is the gravitational potential energy at its maximum value?

Explain.

3. At which point or points is the kinetic energy at its maximum value?

Explain.

4. At which point or points is the total energy at its maximum value?

Explain.

5. At which point or points is the spring potential energy zero?

Explain.

6. At which point or points is the kinetic energy positive?

Explain.

7. At which point or points is the kinetic energy at its maximum value and the gravitational potential energy at its minimum value?

Explain.

NT9D-BCT25: OSCILLATING MASS-SPRING GRAPH OF DISPLACEMENT VS. TIME—ENERGY

A cart attached to a spring is given an initial push, displacing it from its equilibrium position. A graph of displacement as a function of time for the cart is shown at right. The system has a total initial energy of 12 J and there is no friction. Five points are labeled $A - E$ in the graph.

For each labeled point, complete the bar chart below for the kinetic energy and the potential energy for the cart-spring system.

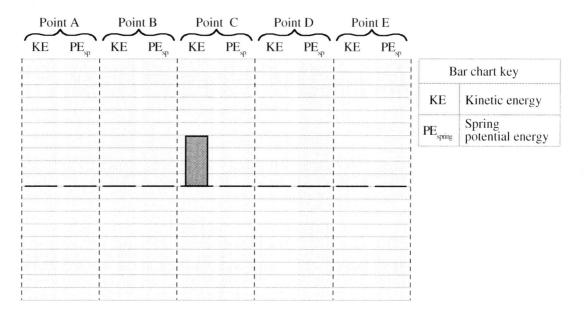

Explain.